SECOND EDITION

CHEMICAL CALCULATIONS
MATHEMATICS FOR CHEMISTRY

SECOND EDITION

CHEMICAL CALCULATIONS

MATHEMATICS FOR CHEMISTRY

PAUL YATES

CRC Press
Taylor & Francis Group
Boca Raton   London   New York

CRC Press is an imprint of the
Taylor & Francis Group, an informa business

CRC Press
Taylor & Francis Group
6000 Broken Sound Parkway NW, Suite 300
Boca Raton, FL 33487-2742

© 2007 by Taylor & Francis Group, LLC
CRC Press is an imprint of Taylor & Francis Group, an Informa business

No claim to original U.S. Government works
Printed in the United States of America on acid-free paper
10 9 8 7 6 5 4 3 2

International Standard Book Number-10: 0-8493-9164-4 (Softcover)
International Standard Book Number-13: 978-0-8493-9164-4 (Softcover)

Library of Congress Cataloging-in-Publication Data

Yates, Paul, 1961-
     Chemical calculations : mathematics for chemistry / Paul Yates. -- 2nd ed.
        p. cm.
     Includes bibliographical references and indexes.
     ISBN-13: 978-0-8493-9164-4 (pbk. : acid-free paper)
     ISBN-10: 0-8493-9164-4 (pbk. : acid-free paper)
     1. Chemistry--Mathematics. I. Title.

QD39.3.M3Y38 2007
540.1'51--dc22
                                                                      2006029548

Visit the Taylor & Francis Web site at
http://www.taylorandfrancis.com

and the CRC Press Web site at
http://www.crcpress.com

For Julie, Catherine, and Christopher

Contents

Preface

It is very gratifying to be able to introduce the second edition of a textbook. The first edition of *Chemical Calculations* clearly addressed a need in chemistry courses, and I believe that this has grown since its first publication. There has also been a growth in the number of textbooks addressing the mathematical skills of chemistry students, so it has been a considerable challenge to try to improve on the first edition while retaining those elements that gave it a distinctive flavor.

My philosophy regarding the overall structure of the book remains the same — it is arranged by chemical rather than mathematical topics. This approach does, I believe, remain unique. I have retained this so that the book can be used throughout a chemistry course long after any specialist module on mathematics for chemists has been completed.

A few chemical topics have been removed where I felt that they did not provide the best means of introducing a mathematical technique. In some cases, however, this deleted material has provided the basis for additional end-of-chapter questions. A number of new topics, both modern and more traditional, have consequently been included. These contain, where appropriate, the spreadsheet treatment of certain topics.

Appendix A in the first edition, which contained the absolutely fundamental aspects of mathematics, has now gained prominence as Chapter 1 at the start of the book. Some students may wish to skip this; the structure is also slightly different from that of subsequent chapters.

One thing that I always look for in any textbook is an array of questions for use by both students and instructors. This aspect has been considerably strengthened in this edition. Questions have been included for the fundamental material in Chapter 1, whereas there were none in Appendix A in the first edition. In the other chapters the numbers of both exercises and chemically related problems have been doubled.

Finally, if you are interested in more general aspects of my activities in chemical education and other material related to that in this book, please have a look at my Web site at www.chemed.org.uk.

Acknowledgments

I would like to thank all the students I have worked with during my career. Certainly, without them there would not have been a first nor second edition of this title. The fact that I can still be asked a question that causes me to rethink my understanding of a topic is a source of great inspiration. Lindsey Hofmeister at Taylor & Francis was instrumental in bringing this edition about and managed to strike the perfect balance between encouragement and freedom. I am indebted to many reviewers of both the published first edition and the proposal for the current edition. I hope that they will forgive me for deciding not to incorporate a suggestion, but please be assured that all were considered!

Textbook writing can be a lonely and sometimes thankless task, so I would like to thank all my colleagues in the chemical education community for their continued friendship and support. I was also delighted to receive an Individual Excellence in Learning and Teaching Award earlier this year, for which I would like to thank the Learning and Teaching Committee at Keele University.

Finally, the greatest thanks must go to my family for their continued belief in me and for their love and support.

Paul Yates
Keele, Staffordshire, United Kingdom

Fundamentals 1

There are some mathematical processes without which you will find it very difficult to solve any problems, no matter how good your grasp of the other concepts involved. These are placed in this preliminary chapter so that you can find them easily if you need to refer to them while studying other parts of the book. You may find it possible to skim over this material, but do have a look at how these concepts are used in real chemical examples.

1.1 Positive and negative numbers

If we have a number represented by the algebraic symbol a, its negative will be written as $-a$. For example, the negative of 4 is -4. There are two rules to remember when combining numbers that may be positive or negative: two like symbols combine to give positive, whereas two unlike symbols combine to give negative. This leads to the following sets of rules.

Use the ⊟ button on your calculator to input a negative number.

1.1.1 Addition

If you are adding a negative number to any other number, this is the same as subtracting the corresponding positive value. For example,

$$2 + (-3) = 2 - 3 = -1$$

because the + and − signs are different and therefore give −. If our two numbers are represented algebraically as a and b, then the rule can be written as

$$a + (-b) = a - b$$

This simple rule finds application in chemistry when we wish to determine the overall change in entropy for a process.

The entropy of a substance gives an indication of the extent of disorder.

This is denoted as $\Delta_{universe}S$, where the term universe is taken to represent both the system of interest and its surroundings. The overall entropy change is then given by

$$\Delta_{universe}S = \Delta_{system}S + \Delta_{surroundings}S$$

where $\Delta_{system}S$ and $\Delta_{surroundings}S$ represent the entropy changes of the system and its surroundings, respectively.

When 2.00 mol of supercooled water at $-15.0°C$ is frozen irreversibly at a constant pressure of 1.00 atm to give ice at $-15.0°C$, the entropy changes in the system and the surroundings are –39.61 and 42.03 J K^{-1}, respectively. The overall entropy change of the universe will then be

$$\Delta_{universe}S = \Delta_{system}S + \Delta_{surroundings}S$$

$$= (-39.61 + 42.03)\ \text{J K}^{-1}$$

$$= 2.42\ \text{J K}^{-1}$$

1.1.2   Subtraction

If you are subtracting a negative number, this is equivalent to adding the corresponding positive number. For example,

$$2 - (-4) = 2 + 4 = 6$$

because we have two like signs combining to give +. Using our symbols a and b we can write the general rule as

$$a - (-b) = a + b$$

We make use of this in chemistry when calculating the expected frequency of a line in an electronic spectrum. The difference ΔE in energy between two levels E_1 and E_2 is given by

$$\Delta E = E_2 - E_1$$

Frequency is converted to energy by multiplying by the Planck constant h.

and ΔE is proportional to the frequency of the resulting line.

The energy levels in an atom are usually measured relative to the zero energy of a free electron, and thus are negative.

In hydrogen the lowest energy level $E_1 = -2.182 \times 10^{-18}$ J and $E_2 = -0.546 \times 10^{-18}$ J. In this case

$$\Delta E = E_2 - E_1$$

$$= -0.546 \times 10^{-18}\ \text{J} - (-2.182 \times 10^{-18}\ \text{J})$$

The adjacent negative signs combine to give a positive, so

$$\Delta E = -0.546 \times 10^{-18}\ \text{J} + 2.182 \times 10^{-18}\ \text{J}$$

From Section 1.1.1 on page 1 this can also be written as

$$\Delta E = 2.182 \times 10^{-18}\ \text{J} - 0.546 \times 10^{-18}\ \text{J}$$

$$= 1.636 \times 10^{-18}\ \text{J}$$

1.1.3   Multiplication

When two positive numbers a and b are multiplied, we obtain another positive number, as in

$$3 \times 4 = 12$$

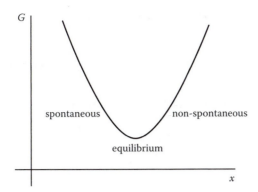

Figure 1.1   Graph of Gibbs free energy G against extent of reaction x. In the spontaneous region of the graph, G is falling, so ΔG is negative, ΔG then becomes zero at equilibrium, and ΔG is positive as G rises in the non-spontaneous region.

We can represent this in symbols as

$$a \times b = ab$$

where it is usual to omit the multiplication sign (×).

If only one of the numbers is negative, then the result will be negative, as in

$$-3 \times 4 = -12, \quad \text{and} \quad 3 \times -4 = -12$$

These are expressed in symbols as

$$-a \times b = -ab \quad \text{and} \quad a \times -b = -ab$$

If, on the other hand, both numbers are negative, then we have two like signs and the result is positive, as shown by

$$-3 \times -4 = 12$$

The general rule is expressed as

$$-a \times -b = ab$$

An area of chemistry where multiplication involving negative numbers occurs is in the calculation of the change ΔG in Gibbs free energy for a process.

The value of ΔG^0 is related to the equilibrium constant K for a process by the equation

$$\Delta G^0 = -RT \ln K$$

where $\ln K$ is the natural logarithm of K. Logarithms will be covered in Chapter 3, Section 3.6.1, on page 75; don't worry about them now, and just think of $\ln K$ as a single number. The other quantities in the equation are the gas constant R and the absolute temperature T.

For the equilibrium

$$2\,HI_{(g)} = H_{2(g)} + I_{2(g)}$$

ΔG is zero at equilibrium, negative for a spontaneous process, and positive for a non-spontaneous process, as shown in Figure 1.1.

R has the value 8.314 J K⁻¹ mol⁻¹.

The absolute temperature is measured in kelvin, K (not °K!). The temperature 0°C is equivalent to 273 K, but a temperature interval of 1°C is the same as a temperature interval of 1 K.

ln K = −3.922 at an absolute temperature 717 K. We thus have

$$\Delta G^0 = -8.314 \text{ J K}^{-1} \text{ mol}^{-1} \times 717 \text{ K} \times (-3.922)$$

Because two negative quantities are being multiplied, the result will be positive, so

$$\Delta G^0 = 8.314 \text{ J K}^{-1} \text{ mol}^{-1} \times 717 \text{ K} \times 3.922$$

$$= 23.4 \times 10^3 \text{ J mol}^{-1}$$

1.1.4   Division

The rules for division are similar to those for multiplication. If 9 is divided by 3, we write

$$9 \div 3 = \frac{9}{3} = 3$$

Similarly, if a is divided by b, we write this as

$$a \div b = \frac{a}{b}$$

If the first number is negative, we have

$$-9 \div 3 = \frac{-9}{3} = -\frac{9}{3} = -3$$

and if the second number is negative we have

$$9 \div (-3) = \frac{9}{-3} = -\frac{9}{3} = -3$$

because we are combining two different signs and therefore obtain a negative result. In general terms, this is written as

$$\frac{-a}{b} = \frac{a}{-b} = -\frac{a}{b}$$

On the other hand, if both numbers are negative, the two like signs give a positive result, as in

$$(-9) \div (-3) = \frac{-9}{-3} = \frac{9}{3} = 3$$

Using our symbols a and b this can be expressed as

$$(-a) \div (-b) = \frac{-a}{-b} = \frac{a}{b}$$

The absolute temperature T at which equilibrium occurs in a system is given by the equation

$$T = \frac{\Delta H}{\Delta S}$$

where ΔH is the change in enthalpy and ΔS is the change in entropy.

For the condensation of hexane we have $\Delta_{condensation} H = -28.9$ kJ mol^{-1}, and $\Delta_{condensation} S = -84.5$ J K^{-1} mol^{-1}. Thus

$$T = \frac{\Delta_{condensation} H}{\Delta_{condensation} S} = \frac{-28.9 \times 10^3 \, \text{J mol}^{-1}}{-84.5 \, \text{J K}^{-1} \, \text{mol}^{-1}} = 342 \, \text{K}$$

Notice the conversion from kJ to J in $\Delta_{condensation} S$. This is necessary to give a consistent set of units with those in $\Delta_{condensation} S$.

1.2 Precedence in equations

It can be difficult to know where to start when evaluating an expression with many terms and mathematical operations. This can be done by applying a set of rules known as the BODMAS rules. BODMAS stands for Brackets of Divide Multiply Add Subtract. These are self explanatory except for "of," which is actually another way of saying "multiply." We also need to realize that raising a number to a power is the same as performing a multiplication. For example

$$2^4 = 2 \times 2 \times 2 \times 2$$

For any expression, we perform the operations in the order specified by the BODMAS rules. Some examples will illustrate this.

Consider the expression

$$2 \times 3 + 4 \times 5$$

The rules tell us to perform multiplication before addition, so, if we do this we have

$$2 \times 3 + 4 \times 5 = 6 + 20$$

which we can evaluate as 26. Now, suppose that the original expression had been modified by the inclusion of brackets to become

$$2 \times (3 + 4) \times 5$$

We now have to evaluate the quantity in brackets first, and as this is equal to 7, we have

$$2 \times (3 + 4) \times 5 = 2 \times 7 \times 5$$

which has the value 70. This example shows us how brackets may be used to override the standard order in which we perform mathematical operations.

An expression that involves raising a number to a power is

$$3 \times 2^5 - 1$$

As noted above, we can write 2^5 as a multiplication, so the expression becomes

$$3 \times 2 \times 2 \times 2 \times 2 \times 2 - 1$$

The BODMAS rules tell us to perform the multiplications first, so the expression becomes

$$96 - 1$$

which has the value 95.

An example of an expression in chemistry that requires application of the BODMAS rules is the van der Waals equation

$$\left(p + \frac{an^2}{V^2} \right)(V - nb) = nRT$$

The amount n is frequently, although incorrectly, known as the "number of moles." The van der Waals equation attempts to model the behaviour of a real gas by including additional correction terms involving the constants a and b to the basic gas equation for an ideal gas.

which relates the pressure p, volume V, and absolute temperature T of an amount n of gas. The quantities a and b are constants for a particular gas.

Let us evaluate the left-hand side of this expression for a sample of 1.00 mol of nitrogen gas occupying a volume of 22.414 dm^3 at a pressure of 1.0122×10^5 Pa. The constants for nitrogen are $a = 0.1408$ Pa m^6 mol^{-2} and $b = 3.913 \times 10^{-5}$ m^3 mol^{-1}.

The first thing that is apparent from the BODMAS rules is that we need to begin with the brackets. In this case it means we evaluate each bracket in turn and then multiply the results together.

Bracket 1: We need to evaluate the term an^2/V^2 containing multiplication (including power) and division operations, and then add this to p.

$$\frac{an^2}{V^2} = \frac{0.1408 \, \mathrm{Pa \, m^6 \, mol^{-2}} \times (1.00 \, \mathrm{mol})^2}{(22.414 \times 10^{-3} \, \mathrm{m^3})^2}$$

$$= \frac{0.1408 \, \mathrm{Pa \, m^6 \, mol^{-2}} \times 1.00 \, \mathrm{mol^2}}{502.39 \times 10^{-6} \, \mathrm{m^6}} = 280.3 \, \mathrm{Pa}$$

$$p + \frac{an^2}{V^2} = 1.0122 \times 10^5 \, \mathrm{Pa} + 280.3 \, \mathrm{Pa} = 1.0150 \times 10^5 \, \mathrm{Pa}$$

Bracket 2: We need to evaluate the term nb containing a multiplication operation, and subtract this from V.

$$nb = 1.00 \, \mathrm{mol} \times 3.913 \times 10^{-5} \, \mathrm{m^3 \, mol^{-1}} = 3.913 \times 10^{-5} \, \mathrm{m^3}$$

$$V = 22.414 \, \mathrm{dm^3} = 22.414 \times 10^{-3} \, \mathrm{m^3}$$

$$V - nb = 22.414 \times 10^{-3} \, \mathrm{m^3} - 3.913 \times 10^{-5} \, \mathrm{m^3} = 2.237 \times 10^{-2} \, \mathrm{m^3}$$

We now multiply the values of each bracket together, so that

$$\left(p + \frac{an^2}{V^2}\right)(V - nb) = 1.0150 \times 10^5 \, \text{Pa} \times 2.237 \times 10^{-2} \, \text{m}^3 = 2.271 \times 10^3 \, \text{Pa} \, \text{m}^3$$

1.3 Rearranging equations

An equation can be thought of as a balance of two sides, each consisting of a series of terms. In order to preserve this balance, whatever operation is applied to one side of the equation must also be applied to the other. This technique can be used to good effect if we wish to rearrange an equation. One reason for doing this might be to isolate one quantity so that its value can be calculated. For example, the equation

$$x + y = 25$$

can be arranged to make y the subject or, in other words, to begin $y = \ldots$ In order to do this, we would need to subtract x from both sides of the equation. This gives

$$x + y - x = 25 - x$$

which simplifies to give

$$y = 25 - x$$

If we substitute actual numbers for x and y we can verify both the original and rearranged expressions. For example, setting $x = 10$ and $y = 15$ gives

$$10 + 15 = 25$$

in the original expression and

$$15 = 25 - 10$$

once it has been rearranged. Clearly, both these statements are correct.
An expression involving multiplication is

$$3y = 2x$$

To rearrange this to the form $y = \ldots$ we need to divide both sides by 3. This gives

$$\frac{3y}{3} = \frac{2x}{3}$$

or

$$y = \frac{2x}{3}$$

If we set $x = 3$ and $y = 2$, then

$$3 \times 2 = 2 \times 3$$

in the original expression and

$$2 = 2 \times \frac{3}{3}$$

once rearranged. Again, both statements are correct.

More complicated expressions contain both addition and subtraction and multiplication and division operations, such as in

$$10x + 5y = 8$$

To obtain an expression of the form $y = ...$ we begin by subtracting $10x$ from either side to give

$$10x + 5y - 10x = 8 - 10x$$

or

$$5y = 8 - 10x$$

We then divide both sides by 5 to give

$$\frac{5y}{5} = \frac{8 - 10x}{5}$$

or

$$y = \frac{8 - 10x}{5}$$

This could also be written as

$$y = \frac{8}{5} - 2x$$

In chemistry the heat capacity C_p is frequently given as a series of terms of increasing power in the variable T, the absolute temperature. The simplest representation of this is

$$C_p = a + bT$$

where a and b are constants. It is possible to rearrange this to give T as a function of C_p. Subtracting a from both sides gives

$$C_p - a = a + bT - a$$

or

$$C_p - a = bT$$

Dividing both sides by b gives

$$\frac{C_p - a}{b} = \frac{bT}{b}$$

or

$$\frac{C_p - a}{b} = T$$

Conventionally, we would now reverse the order of the equation to give

$$T = \frac{C_p - a}{b}$$

although this does not change its meaning in any way.

1.4   Fractions

A fractional quantity is written as two numbers separated by a horizontal line. The fraction 4/5 means "divide 4 by 5." This can also be written using the division sign as $4 \div 5$ and has the decimal value 0.8. To summarize,

$$4 \div 5 = \frac{4}{5} = 0.8$$

Using algebraic symbols

$$\frac{a}{b}$$

means divide a by b, and may also be written as $a \div b$.

1.4.1   Identical fractions

If both the top and bottom of a fraction are multiplied by a constant, its value remains unchanged:

$$\frac{1}{2} = \frac{2 \times 1}{2 \times 2} = \frac{2}{4}$$

Using algebraic symbols for the general case:

$$\frac{a}{b} = \frac{c \times a}{c \times b} = \frac{ca}{cb}$$

Similarly, we can divide both top and bottom by the same number and the value of the fraction remains unchanged:

$$\frac{3}{9} = \frac{3/3}{9/3} = \frac{1}{3}$$

Again, using symbols to represent a general case:

$$\frac{ad}{bd} = \frac{ad/d}{bd/d} = \frac{a}{b}.$$

This technique is frequently used when we perform the stoichiometric determination of an empirical formula. Such a determination of a compound containing only phosphorus and oxygen showed that a sample contained 1.409 mol of phosphorus (P) and 3.523 mol of oxygen (O). The ratio of these quantities can be expressed as the fraction

$$\frac{P}{O} = \frac{1.409}{3.523}$$

The top and bottom of this fraction can be divided by 1.409 to give

$$\frac{P}{O} = \frac{1.409/1.409}{3.523/1.409} = \frac{1}{2.5}$$

The top and bottom of this can each be multiplied by 2 to give

$$\frac{P}{O} = \frac{1}{2.5} = \frac{1 \times 2}{2.5 \times 2} = \frac{2}{5}$$

and hence the empirical formula of the compound is P_2O_5.

1.4.2 Addition and subtraction

If we wish to add two fractions, or to subtract one fraction from another, we need to rewrite them in terms of what is called a common denominator. This means that the number on the bottom of each of the two fractions needs to be the same. For example, we cannot calculate the sum

$$\frac{2}{3} + \frac{1}{4}$$

until the two fractions are rewritten in terms of a common denominator. The common denominator in this case is 12, obtained by multiplying 3 and 4. We can write

$$\frac{2}{3} = \frac{2 \times 4}{3 \times 4} = \frac{8}{12}$$

and

$$\frac{1}{4} = \frac{1 \times 3}{4 \times 3} = \frac{3}{12}$$

The sum is now easily evaluated as

$$\frac{2}{3} + \frac{1}{4} = \frac{8}{12} + \frac{3}{12} = \frac{11}{12}$$

The addition of two general fractions can be expressed as

$$\frac{a}{b} + \frac{c}{d} = \frac{ad}{bd} + \frac{bc}{bd} = \frac{ad + bc}{bd}$$

It is worth noting that it may be possible to simplify the result further, by dividing both top and bottom by the same number, as outlined previously in Section 1.4.1 on page 9.

The van der Waals equation, met in Section 1.2 on page 5, can be rearranged to

$$p = \frac{nRT}{V - nb} - \frac{an^2}{V^2}$$

The common denominator of the two fractions on the right-hand side will be

$$V^2(V - nb)$$

We then have

$$\frac{nRT}{V - nb} = \frac{nRTV^2}{V^2(V - nb)}$$

and

$$\frac{an^2}{V^2} = \frac{an^2(V - nb)}{V^2(V - nb)}$$

The overall equation can then be written as

$$p = \frac{nRTV^2 - an^2(V - nb)}{V^2(V - nb)}$$

1.4.3   Multiplication

When two fractions are multiplied, we multiply the terms on the top and the terms on the bottom independently. For example,

$$\frac{2}{3} \times \frac{1}{4} = \frac{2 \times 1}{3 \times 4} = \frac{2}{12}$$

In this case we can divide both top and bottom by 2, so that

$$\frac{2}{3} \times \frac{1}{4} = \frac{1}{6}$$

In general terms, using symbols,

$$\frac{a}{b} \times \frac{c}{d} = \frac{ac}{bd}$$

When the product in a chemical reaction becomes a reactant in a second chemical reaction, the two reactions are said to be coupled. Such a pair of reactions is

$$PO_4^{3-} + H_3O^+ = HPO_4^{2-} + H_2O$$

$$HPO_4^{2-} + H_3O^+ = H_2PO_4^- + H_2O$$

which can be combined to give the overall reaction

$$PO_4^{3-} + 2\,H_3O^+ = H_2PO_4^- + 2\,H_2O$$

The product HPO_4^{2-} in the first reaction is a reactant in the second reaction. The equilibrium constant K_1 for the first reaction is given by

$$K_1 = \frac{[HPO_4^{2-}][H_2O]}{[PO_4^{3-}][H_3O^+]}$$

and that for the second reaction, K_2, is given by

$$K_2 = \frac{[H_2PO_4^-][H_2O]}{[HPO_4^{2-}][H_3O^+]}$$

If K_1 and K_2 are multiplied, they give the equilibrium constant K for the overall reaction

$$K = K_1 K_2 = \frac{[HPO_4^{2-}][H_2O]}{[PO_4^{3-}][H_3O^+]} \times \frac{[H_2PO_4^-][H_2O]}{[HPO_4^{2-}][H_3O^+]}$$

The term $[HPO_4^{2-}]$ appears on both the top and bottom of this expression, and so can be cancelled to give the overall equilibrium constant as

$$K = \frac{[H_2PO_4^-][H_2O]^2}{[PO_4^{3-}][H_3O^+]^2}$$

because $[H_2O] \times [H_2O] = [H_2O]^2$, and $[H_3O^+] \times [H_3O^+] = [H_3O^+]^2$.

1.4.4   Division

When a number is divided by a fraction, the operation we perform is to multiply by the reciprocal of the fraction. The reciprocal of a fraction is the fraction turned upside down. For example,

$$\frac{2}{3} \div \frac{1}{4} = \frac{2}{3} \times \frac{4}{1} = \frac{2 \times 4}{3 \times 1} = \frac{8}{3}$$

In symbols, for the general case

$$\frac{a}{b} \div \frac{c}{d} = \frac{a}{b} \times \frac{d}{c} = \frac{ad}{bc}$$

The example in Section 1.4.1 on page 9 introduced the use of fractions in the determination of the empirical formula in a compound containing only phosphorus and oxygen. We can generalize this and relate it to the mass m of each element and its atomic mass M.

The amount n_P of phosphorus will be given by

$$n_P = \frac{m_P}{M_P}$$

where m_P is the mass of phosphorus in the compound and M_P is the atomic mass of phosphorus. In a similar fashion the amount n_O of oxygen is given by

$$n_O = \frac{m_O}{M_O}$$

where m_O is the mass of oxygen in the compound, and M_O is the atomic mass of oxygen.

The ratio of the amounts of phosphorus and oxygen is then given by

$$\frac{n_P}{n_O} = \frac{m_P/M_P}{m_O/M_O} = \frac{m_P}{M_P} \times \frac{M_O}{m_O}$$

using the rule that dividing by a fraction is equivalent to multiplying by its reciprocal.

1.5   Indices

A quantity such as 2^4 means that the base 2 is raised to the power 4 (or has an index of 4), and

$$2^4 = 2 \times 2 \times 2 \times 2$$

Here we have multiplied 2 together 4 times. In general x^n is the base x raised to the power n, and consists of x multiplied together n times.

Some frequently encountered powers are given special names:

x^2 is called "x squared" and is equal to $x \times x$
x^3 is called "x cubed" and is equal to $x \times x \times x$

A little thought will show that $x^1 = x$, and we also have the special case that $x^0 = 1$.

There are a number of rules for combining numbers raised to powers. It is important to remember that such rules only apply when the quantities being combined have the same base.

1.5.1   Multiplication

When two numbers with the same base raised to powers are multiplied, we add the powers so that

$$2^2 \times 2^3 = 2^{(2+3)} = 2^5$$

In general terms, this can be written as

$$x^a \times x^b = x^{(a+b)}$$

The ionic product K_w of water is given by

$$K_w = [H^+][OH^-]$$

In the case of pure water, $[H^+] = [OH^-]$, so the expression becomes

$$K_w = [H^+][H^+]$$

Because $[H^+] = [H^+]^1$, this can be written as

$$K_w = [H^+]^1 \times [H^+]^1$$

$$= [H^+]^{(1+1)} = [H^+]^2$$

A similar argument can be used to show that we could alternatively write

$$K_w = [OH^-]^2$$

1.5.2   Division

When one number raised to a power is divided by a number consisting of the same base raised to a power, we take the difference between the powers, so that, for example,

$$\frac{2^5}{2^3} = 2^{(5-3)} = 2^2$$

In general terms,

$$\frac{x^a}{x^b} = x^{(a-b)}$$

The equilibrium constant K for the dissociation of chlorine

$$Cl_2 = 2\,Cl$$

is given by

$$K = \frac{[Cl]^2}{[Cl_2]}$$

At the point in the dissociation where

$$[Cl] = [Cl_2]$$

$$K = \frac{[Cl]^2}{[Cl]}$$

Applying the rules of indices then gives

$$K = [Cl]^{(2-1)} = [Cl]$$

At this point it will also be true that

$$K = [Cl_2]$$

1.5.3   Raising to a power

When a number raised to a power is itself raised to a power, we multiply the powers, as in

$$(2^2)^3 = 2^{(2 \times 3)} = 2^6$$

This can be expressed in general terms as

$$(x^a)^b = x^{(ab)}$$

Use the calculator button to raise a number to a power.

Copper(II) iodate is sparingly soluble in water according to the equation

$$Cu(IO_3)_{2(s)} = Cu^{2+}_{(aq)} + 2\,IO3^-_{(aq)}$$

If the small concentration of Cu^{2+} in solution is x, it follows from the stoichiometry of the equation that the concentration of IO_3^- is $2x$. The solubility product K_{sp} is defined as

$$K_{sp} = [Cu^{2+}][IO_3^-]_2$$
$$= x\,(2x)^2$$
$$= x \times 4x^2$$
$$= 4x^3$$

Use the button to take the square root of a number.

1.5.4   Roots

The square root of the number 16 is 4. This means that 4 multiplied by itself gives 16.

We write this mathematically as

$$\sqrt{16} = 4$$

or using an index as

$$16^{\frac{1}{2}} = 4$$

It is also possible to calculate other roots so that the root raised to the appropriate power gives back the original number. For example, we can write that

$$\sqrt[3]{8} = 2$$

which states that the cube root of 8 is 2, and it follows that

$$2^3 = 8$$

Another way of expressing the cube root of 8 is as

$$\sqrt[3]{8} = 8^{\frac{1}{3}}$$

and, in general, when we take the root of degree b of a number x raised to power a, the index of x is a divided by b, so that

$$\sqrt[b]{x^a} = x^{\frac{a}{b}}$$

Quantities raised to fractional indices occur frequently in kinetics. An example is the reaction

$$CH_3CHO \rightarrow CH_4 + CO$$

for which the rate of reaction is equal to $\left[CH_3CHO \right]^{\frac{3}{2}}$.

1.5.5   Negative powers

A negative power denotes a reciprocal quantity, i.e., "one over" the number. For example

$$2^{-3} = \frac{1}{2^3}$$

In general algebraic terms

$$x^{-n} = \frac{1}{x^n}$$

For the equilibrium dissociation of hydrogen chloride

$$2\,HCl = H_2 + Cl_2$$

the equilibrium constant K will be given by

$$K = \frac{[H_2][Cl_2]}{[HCl]^2}$$

Because

$$\frac{1}{x^2} = x^{-2}$$

this can be rewritten as

$$K = [H_2][Cl_2][HCl]^{-2}$$

1.6 Standard form

Any number can be written as the product of a number between 1 and 10 and the number 10 raised to an integral power. For example,

$$2907 = 2.907 \times 10^3$$

$$0.000\ 46 = 4.6 \times 10^{-4}$$

Notice that for a number less than 1 the power of 10 will be negative. To convert from standard form we note that a negative power, 10^{-n}, means move the decimal point n places to the left, whereas a positive power, 10^n, means move the decimal point n places to the right. In both cases we need to insert an appropriate number of zeros to act as place holders.

For the reaction

$$SO_2Cl_{2(g)} \rightarrow SO_{2(g)} + Cl_{2(g)}$$

the rate constant k has the value 2.20×10^{-5} s^{-1}. Moving the decimal point 5 places to the left gives 0.0000220 s^{-1} once the necessary zeros are inserted. The half-life $t_{1/2}$ for this reaction is 3.15×10^4 s. The positive value of the power indicates that we move the decimal place 4 places to the right; this gives 31500 s.

We use the EXP button to enter powers of 10.

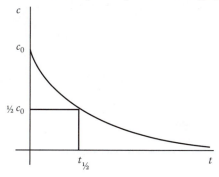

The half-life of a reaction is the time taken for the concentration of reactant to fall to half its initial value, as shown in Figure 1.2.

Figure 1.2 Graph of concentration c against time t for a reaction. The time taken for the concentration to fall to half its initial value c_0 is the half-life denoted by $t_{1/2}$.

Exercises

1. Calculate

    (a) $3 + (-1)$ (b) $-7 + 2$ (c) $-4 + (-2)$

2. Calculate

    (a) $4 - 6$ (b) $5 - (-2)$ (c) $-6 - (-1)$

3. Calculate

    (a) 4×7 (b) $6 \times (-2)$ (c) $-3 \times (-7)$

4. Calculate

    (a) $\dfrac{28}{4}$ (b) $\dfrac{21}{-3}$ (c) $\dfrac{-81}{-9}$

5. Evaluate

    (a) $3^2 - 18/3 + 2 \times 3$

    (b) $(3^2 - 18)/(3 + 2 \times 3)$

    (c) $3^2 - 18/(3 + 2 \times 3)$

6. Rearrange the following equations to make y the subject, i.e., so that they begin with $y = \ldots$

    (a) $x - y = 10$

    (b) $3x + y = 8$

    (c) $4x - 2y = 10$

7. Simplify each of the following to a single term

    (a) $3^4 \times 3^2$

    (b) $x^2 \times x^4$

    (c) $x^3 \times x^{-2}$

8. Simplify each of the following to a single term

    (a) $6^3 \div 6^4$

    (b) $x^2 \div x^3$

    (c) $y^4 \div y^{-2}$

9. Simplify each of the following to a single term

   (a) $(4^2)^6$

   (b) $(x^3)^4$

   (c) $(y^{-2})^6$

10. Write each of the following in terms of a fractional index

    (a) $\sqrt{3^4}$ (b) $\sqrt[3]{x^5}$ (c) $\sqrt[5]{y^{-3}}$

11. Write the following in the form ax^n

    (a) $\dfrac{1}{4^3}$ (b) $\dfrac{3}{x^5}$ (c) $\dfrac{2}{y^{-2}}$

12. Simplify the following fractions

    (a) $\dfrac{12}{15}$ (b) $\dfrac{6}{9}$ (c) $\dfrac{8x}{12y}$

13. Determine the following in their simplest fractional form

    (a) $\dfrac{4}{7}+\dfrac{3}{4}$ (b) $\dfrac{7}{8}+\dfrac{3}{4}$ (c) $\dfrac{2}{3}-\dfrac{5}{9}$

14. Determine the following in their simplest fractional form

    (a) $\dfrac{7}{8}\times\dfrac{3}{4}$ (b) $\dfrac{2}{3}\times\dfrac{8}{9}$ (c) $\dfrac{5}{6}\times\dfrac{2}{7}$

15. Determine the following in their simplest fractional form

    (a) $\dfrac{3}{4}\div\dfrac{5}{6}$ (b) $\dfrac{2}{3}\div\dfrac{8}{9}$ (c) $\dfrac{3}{5}\div\dfrac{2}{3}$

16. Express the following numbers in standard form

    (a) 36387   (b) 2238000   (c) 0.000 176

17. Rewrite the following numbers without using a power of 10

    (a) 3.62×10^7 (b) 1.66×10^4 (c) 1.27×10^{-5}

18. Calculate

    (a) $\dfrac{1}{2}\times\left(-\dfrac{3}{8}\right)$ (b) $\dfrac{7}{8}-\left(-\dfrac{8}{9}\right)$ (c) $\dfrac{2}{3}\div\left(-\dfrac{3}{4}\right)$

19. Evaluate

   (a) $2^3 - 6/(-2) + 3 \times (-4)$

   (b) $4^2 + (-12)/(-3) - 2 \times (-6)$

   (c) $(-5)^3 - (-8)/(-2) - 4 \times (-3)$

20. Evaluate

   (a) $\left(-\dfrac{1}{2}\right)^2 + \left(-\dfrac{3}{8}\right) - \dfrac{2}{3} \times \left(-\dfrac{1}{2}\right)$

   (b) $-\left(\dfrac{3}{4}\right)^2 - \left(-\dfrac{5}{8}\right) \div \left(-\dfrac{3}{4}\right)$

   (c) $\left(-\dfrac{2}{3}\right)^2 - \left(-\dfrac{1}{4}\right)^2 \div \left(-\dfrac{1}{2}\right)$

Problems

1. The Born–Haber cycle for sodium chloride allows its lattice enthalpy $\Delta_L H$ to be calculated in terms of various other enthalpy changes. This leads to the equation

   $$147 \text{ kJ mol}^{-1} - (-639 \text{ kJ mol}^{-1}) = -(-493 \text{ kJ mol}^{-1})$$
   $$- (-293 \text{ kJ mol}^{-1}) - (-\Delta_L H)$$

   Calculate $\Delta_L H$ from this equation.

2. Coulomb's law gives the potential energy $V(r)$ in terms of the distance r between two point charges q_1 and q_2 as

   $$V(r) = \frac{q_1 q_2}{4\pi\varepsilon_0 r}$$

   where ε_0 is a constant known as the permittivity of free space. Determine an expression for the potential energy $V(r)$ between a nucleus with atomic number Z having charge Ze and an electron having charge $-e$.

3. For a chemical reaction the change in Gibbs free energy ΔG is related to the change in enthalpy ΔH and the change in entropy ΔS by

   $$\Delta G = \Delta H - T\Delta S$$

   where T is the absolute temperature. Rearrange this equation to give an expression for T.

4. The equation for catalytic methanation is

$$CO_{(g)} + 3\,H_{2(g)} = CH_{4(g)} + H_2O_{(g)}$$

with a corresponding equilibrium constant K given by

$$K = \frac{[CH_4][H_2O]}{[CO][H_2]^3}$$

Determine a simplified form of this expression when

(a) $[H_2] = [H_2O]$ (b) $[H_2] = 2\,[CO]$

(c) $[H_2] = \frac{1}{2}\,[CO]$ (d) $[CO] = [H_2]$

5. The rate equation for the reaction

$$2\,NO_{2(g)} \rightarrow 2\,NO_{(g)} + O_{2(g)}$$

is

$$\frac{1}{[NO_2]} = kt + \frac{1}{[NO_2]_0}$$

where $[NO_2]_0$ is the initial concentration of NO_2, and $[NO_2]$ its concentration after time t. The rate constant k is a measure of how fast the reaction proceeds.

Rearrange this equation to express

(a) kt as a difference of two fractions
(b) kt in terms of a single fraction with denominator $[NO_2][NO_2]_0$.

6. The quantum yield ϕ for a photochemical reaction can be expressed in terms of the rate constants k for various processes as

$$\frac{1}{\phi} = \frac{k_F + k_{IC} + k_{ISC}}{k_{ISC}}\left(1 + \frac{k_P + k'_{ISC}}{k_R[R]}\right)$$

where $[R]$ is the concentration of reactant. The processes involved are fluorescence (F), internal conversion (IC) and intersystem crossing (ISC), and k_R is the bimolecular rate constant. Write an expression for $1/\phi$ as a single fraction.

7. Write the following physical constants and conversion factors in standard form:

(a) $c = 299\,800\,000\ \text{m s}^{-1}$

(b) $R = 10\,970\,000\ \text{m}^{-1}$

(c) $F = 96\,490\ \text{C mol}^{-1}$

(d) $p = 101\ 325$ Pa

(e) $l = 1\ 609\ 244$ mm

8. The wavenumbers $\bar{\upsilon}$ of lines in the hydrogen-like atom spectrum are given by

$$\bar{\upsilon} = RZ^2\left(\frac{1}{n_1^2} - \frac{1}{n_2^2}\right)$$

where n_1 is the lower level of the transition, n_2 the upper level, R the Rydberg constant, and Z the nuclear charge. Obtain expressions for $\bar{\upsilon}$ when

(a) $n_1 = 1$ and $n_2 = 2$

(b) $n_1 = 2$ and $n_2 = 3$

(c) $n_2 = 2n_1$

9. The heat capacity C_p of a substance at constant pressure can be expressed in terms of the absolute temperature T as

$$C_p = a + bT + \frac{c}{T^2}$$

where a, b, and c are constants. For graphite the constants take the values $a = 16.86$ J K^{-1} mol^{-1}, $b = 4.77 \times 10^{-3}$ J K^{-2} mol^{-1}, and $c = -8.54 \times 10^5$ J K mol^{-1}. Determine the value of C_p for graphite at 325 K.

10. The decomposition of ozone catalyzed by N_2O_5 is represented by the equation

$$2\,O_3 = 3\,O_2$$

and the rate v of this reaction is given by

$$v = k[O_3]^{\frac{2}{3}}[N_2O_5]^{\frac{2}{3}}$$

where k is the rate constant.

(a)   Obtain an expression for the rate when $[O_3] = [N_2O_5]$.

(b)   Obtain an expression for the rate when $[O_3] = 2\,[N_2O_5]$

(c)   Obtain an expression for the rate when $[O_3] = 0.12$ mol dm^{-3}, and $[N_2O_5] = 0.16$ mol dm^{-3}

Dealing with uncertainty in experimental techniques 2

Despite the increasing use of computers in many aspects of chemistry and the enormous advances in theoretical areas such as quantum mechanics, most chemists still perform laboratory experiments. Particularly in physical chemistry, the object of an experiment is often to collect some form of numerical data that is then analyzed to determine the value of some physical quantity. It is important that you are able to handle this type of data. At first sight this may seem trivial, but in some ways the appropriate treatment of data is one of the most difficult areas of mathematical chemistry to master. Every case seems to be different, and it is only by gaining experience of several calculations that you will become confident in performing this kind of data analysis.

2.1 Displaying uncertainties in the measurement of absorbance

We will begin our consideration of experimental uncertainties by considering an example of how these are expressed.

A solution having a concentration of 0.002 10 mol dm^{-3} of bromine in tetrachloromethane was found to have an absorbance of 0.796 ± 0.003. The presence of the plus and minus sign $\pm$ in this expression immediately alerts us to the fact that we are being presented with a measured quantity together with its uncertainty. Taking the minus operation first suggests that

$$\text{Minimum absorbance} = 0.796 - 0.003 = 0.793$$

Whereas taking the addition operation suggests that

$$\text{Maximum absorbance} = 0.796 + 0.003 = 0.799$$

Taking these values together, it is reasonable to state that we would expect the absorbance of this sample to fall between 0.793 and 0.799. As we will see, there are different ways of determining uncertainties, so the precise meaning of a statement like this will be slightly vague at the moment.

We have used the term uncertainty in this discussion; you will also see it called an experimental error, or just error. Because the term error could suggest a mistake, although uncertainties are necessarily associated with all experimental measurements, we will continue to use the term uncertainty.

Accuracy and precision are terms that are often confused. Accuracy denotes how close a measured value is to the true value of a quantity, whereas precision is an indication of how reliably a measurement can be reproduced. It is thus possible to have precise measurements that are not accurate, and accurate measurements that are not precise.

2.2   Measurement in chemistry

Many of the measurements made in physical chemistry are of quantities that are easy to visualize. For example, a burette is used to measure volume, and a thermometer is used to measure temperature. Other measurements that you may make in a university chemistry course will be of less familiar quantities. Absorbance is measured using a spectrophotometer, and electromotive force is measured using a digital voltmeter, for example. Frequently, the values read from these instruments will be used in subsequent calculations, which will usually be performed with the assistance of an electronic calculator. Such calculators do allow you to perform complex calculations easily, but unfortunately they do not always give results that can be quoted directly to give an appropriate answer.

A common example will serve to illustrate this, using a procedure you will have already encountered in the laboratory. Suppose you wish to standardize a solution of sodium hydroxide that has a concentration of approximately 0.1 mol dm^{-3}. To do this, titrate 25 cm^3 of the sodium hydroxide with hydrochloric acid, which, you are told, has a concentration of 0.0994 mol dm^{-3}. Suppose that 25 cm^3 of the sodium hydroxide is neutralized by 24.85 cm^3 of the hydrochloric acid. The concentration of sodium hydroxide, [NaOH], is then given by the formula

$$[NaOH] = \frac{24.85 \, cm^3 \times 0.0994 \, mol \, dm^{-3}}{25 \, cm^3}$$

Using the calculator, this gives a value for [NaOH] of 0.0988036 mol dm^{-3}. The question we need to ask ourselves is, "Are we justified in quoting the final answer to this many figures?" You may guess that, in this case, the answer is "no" because there is clearly a disparity between the number of figures in this answer and that in the original data. This is a vague statement, however, and how do we decide what number of figures would be reasonable in the final answer? To do this we need to understand the concepts of decimal places and significant figures.

2.2.1   Decimal places

In the example above, we could quote the final value for [NaOH] to any number up to 7 decimal places. Some examples would be

[NaOH] = 0.0988 mol dm^{-3} to 4 decimal places
[NaOH] = 0.098 80 mol dm^{-3} to 5 decimal places
[NaOH] = 0.098 803 6 mol dm^{-3} to 7 decimal places

The number of decimal places is the number of digits that appear after the decimal point.

2.2.2   Significant figures

The number of significant figures in each expression is, however, quite different from the number of decimal places. To count these, we need to apply the following rules:

1. Ignore any leading zeros (such as those in 0.001, for example).
2. Include any zeros "inside" the number (such as the second zero in 0.101, for example).
3. Use your judgment to decide whether trailing zeros are significant or not. This is more likely to be the case when they appear after the decimal point. If a number has been expressed as 37.00, for example, it is unlikely that the writers really intended to write the more approximate 37, at least if they had thought what they were doing! Similarly, writing 0.003 700 with four significant figures implies greater precision than does writing 0.0037 with only two.
4. All nonzero digits are significant.

So, for the expressions of sodium hydroxide concentration above, we now have:

[NaOH] = 0.0988 mol dm^{-3} to 3 significant figures because the leading zeros are not significant.
[NaOH] = 0.098 80 mol dm^{-3} to 4 significant figures, since the leading zeros are not significant and the trailing zero is significant as it increases the precision to which the number is specified.
[NaOH] = 0.098 803 6 mol dm^{-3} to 6 significant figures because the leading zeros are not significant, and the enclosed zero is significant.

Now, consider what happens if we wish to express the concentration to 5 significant figures. We might expect to obtain a value of 0.098 803 mol dm^{-3}, but if we look at the first digit immediately to the right of this, which we have deleted, we find that it is a 6. This means that the value we require is, in fact, closer to 0.098 804 mol dm^{-3}, and we need to "round up" the last digit to give the higher value.

When quoting a value to a specified number of decimal places or significant figures, always look at the first digit you are planning to delete. If it is between 5 and 9, round up the previous digit; otherwise leave the previous digit as it is.

One case in which the application of the rules of significant figures can seem a little strange is when you are asked to round a number greater than 10 to fewer figures than there are digits present before the decimal point, such as

1265 to 2 significant figures. In such a case, we need to fill out the remaining places with zeros; in this case we need to round up, successively, 1265 rounds to 1270, which in turn rounds to 1300. Remember that subsequently the zeros in this number are not significant.

Worked example 2.1

Write each of the following quantities to (i) 3 significant figures and (ii) 2 decimal places:

(a) a volume of 17.927 cm^3
(b) an energy change of 396.748 kJ mol^{-1}
(c) a rate constant of 1.0846 dm^3 mol^{-1} s^{-1}
(d) the gas constant 0.082 057 8 dm^3 atm K^{-1} mol^{-1}

Chemical background

(a) We have already seen that measurement of volume is important in chemistry in volumetric analysis. Volume changes are also important in thermodynamics where much emphasis is placed on the behaviour of gases.
(b) Energy changes accompany chemical reactions and can be predicted using thermodynamic methods. They can be measured experimentally using one of various types of calorimeters available, such as that shown in Figure 2.1.
(c) The rate constant determines how fast a particular chemical reaction will take place. You may have noticed that the units can be thought of as "per concentration per time," and it is the value of the concentration that often governs the rate of a chemical reaction at a given temperature. As we will see in Chapter 5, the rate constant has different units depending on the type of reaction to which it refers.
(d) We usually express the gas constant in units of J K^{-1} mol^{-1}, but the form given in this example can be useful when pressure values are given in units of atm. The gas constant appears in many branches of physical chemistry and not just those concerned with gases!

Figure 2.1   Schematic diagram of a Dewar calorimeter.

Solution to worked example 2.1

(a) Counting the first 3 significant figures is straightforward and gives a value of 17.9 cm^3. To 2 decimal places, we obtain 17.92 cm^3, but a 7 has been deleted (which falls between 5 and 9), so we round up the last digit to give 17.93 cm^3.

(b) The first 3 significant figures take us as far as the decimal point, but the first digit deleted is 7 (falling between 5 and 9), so we round 6 up to 7 to give 397 kJ mol^{-1}. To 2 decimal places we have 396.74 kJ mol^{-1}, but the first digit deleted is 8 so we round up to get 396.75 kJ mol^{-1}.

(c) The first 3 significant figures give 1.08 dm^3 mol^{-1} s^{-1} because the zero inside the number is included as a significant digit. The first digit deleted is 4 (which falls between 0 and 4), so no rounding is required. The result is the same when we round to 2 decimal places.

(d) We ignore the leading zeros and count 3 significant figures to give 0.0820 dm^3 atm K^{-1} mol^{-1}. However, the first digit deleted is 5, so we round up to give 0.0821 dm^3 atm K^{-1} mol^{-1}. Counting 2 decimal places gives 0.08 dm^3 atm K^{-1} mol^{-1}, and as a 2 has been deleted (between 0 and 4), no additional rounding is required.

2.2.3   Combining quantities

Having decided how many significant figures are appropriate for each quantity involved in a calculation, we are in a position to determine the appropriate number of digits to which to quote our final calculated value. To do this we need to apply the following rules:

1. When adding or subtracting numbers, the answer should be given to the smallest number of decimal places. For example, we would write $1.23 + 4.5 = 5.7$ rather than $1.23 + 4.5 = 5.73$.

2. When multiplying numbers, or dividing one number by another, the answer should be given to the smallest number of significant figures. For example, we would write

$$\frac{7.432}{2.1} = 3.5$$

rather than

$$\frac{7.432}{2.1} = 3.539$$

Worked example 2.2

Find the value of the following expressions, giving the appropriate number of figures in your answer.

(a) The difference in burette readings of 36.35 cm³ and 11.2 cm³

(b) The density of a solution calculated from the formula

$$\text{density} = \frac{\text{mass}}{\text{volume}} = \frac{17.098\,\text{g}}{6.2\,\text{cm}^3}$$

(c) The concentration of a solution given by the expression

$$\frac{25.15\,\text{cm}^3 \times 0.104\,\text{mol dm}^{-3}}{25.00\,\text{cm}^3}$$

Chemical background

(a) This is actually an example of bad laboratory practice. The burette should always be read to the same degree of precision, which would typically be to ±0.05 cm³. This is true even if the starting value was exactly zero, so this would be given as 0.00 cm³ rather than 0.0 cm³.

(b) The density of a substance is found by dividing the mass of a sample by its volume. One example of doing this in the laboratory, for solutions, is an experiment to determine partial molar volumes.

(c) This is another example of the standardization procedure met in Section 2.2. It is important to realize that the division by 25.00 cm³ refers to a pipette volume, so the trailing zeros are important to take account of the precision of the pipette.

Solution to worked example 2.2

(a) The subtraction (36.35 − 11.2)cm³ on a calculator gives the result 25.15 cm³. However, we can see that the second of the two values is given to only one decimal place, so the final answer would be 25.1 cm³. However, the digit 5 has been deleted, so we need to round up the final value to give 25.2 cm³.

(b) Using the calculator gives an answer of 2.757 741 9 g cm⁻³, which contains more figures than we require. If we look at the two quantities given in the question, we see that they contain 5 and 2 significant figures, respectively. Giving the answer to 2 significant figures gives 2.7 g cm⁻³, but we need to remember that we have deleted the digit 5, so we need to round up the answer to 2.8 g cm⁻³.

(c) The calculator again gives a large number of figures, and we obtain an initial answer of 0.104 624 mol dm⁻³. Looking at the three quantities in the question shows that they are given to 4, 3, and 4 significant figures, respectively. We include in this count the zero inside the second quantity and the trailing zeros in the pipette volume. The answer therefore needs to be given to 3 significant figures, which gives 0.104 mol dm⁻³. The first digit deleted is 6, so, once again, we need to round up so that the final answer is given as 0.105 mol dm⁻³.

2.3   Stoichiometric calculations

In the previous section, we used calculations based on volumetric analysis as examples. The reaction between sodium hydroxide and hydrochloric acid takes place in a simple 1:1 ratio, so we did not have to consider this aspect of the reaction any further. Many chemical reactions are less straightforward than this, but it is useful to be able to calculate the amount or mass of products from a given amount or mass of reactants.

2.3.1   Multiplication and division by an integer

In previous examples of calculations we have been dealing with real numbers where it has been necessary to determine the number of significant figures (or decimal places) in each value included. However, when multiplying or dividing by an integer, we do not need to consider how many significant figures it has. This is because integers are *exact* numbers, so that 4 could be thought of as 4.0000000, ..., for example, to ensure that it is treated as having more significant figures than any other quantity in the calculation.

Some simple examples of this are

$$3 \times 5.983 = 17.95 \text{ given to the 4 significant figures of } 5.983$$

and

$$\frac{3.16}{5} = 0.632 \text{ given to the 3 significant figures of } 3.16.$$

Worked example 2.3

In the reaction

$$Fe_2O_{3(s)} + 3\,CO_{(g)} \rightarrow 2\,Fe_{(l)} + 3\,CO_{2(g)}$$

calculate the mass of iron produced when 503 g of carbon monoxide is passed over an excess of Fe_2O_3. The molecular mass of CO is 28.01 g mol^{-1}, and the atomic mass of iron is 55.85 g mol^{-1}.

Chemical background

This equation represents the reaction that takes place in a blast furnace (Figure 2.2), where iron is extracted from iron ore by reduction with carbon monoxide. Coke is burned to form carbon dioxide, which then reacts with more hot coke to form carbon monoxide.

Unlike the previous examples we have seen, where an equal number of molecules of each reactant combine, here we see that 1 molecule of iron oxide combines with 3 molecules of carbon monoxide. The fact that these numbers are still integers is a consequence of Dalton's atomic theory expressed in the law of multiple proportions, which was formulated in 1803.

A discussion of the significance of Dalton's atomic theory can be found in *Chemistry in Britain*, **30**, 920–921, 1994.

Figure 2.2   Schematic diagram of a blast furnace.

Solution to worked example 2.3

It is important to realize that the coefficients that appear in this equation refer to the relative numbers that combine in the reaction. These are known as the amounts of each species and are measured in moles. Using the relationship

$$amount = \frac{mass}{molecular\ mass}$$

and substituting the given values into this equation, we have

$$amount\ of\ CO = \frac{503\ g}{28.01\ g\ mol^{-1}}$$

and

$$amount\ of\ Fe = \frac{mass\ of\ Fe}{55.85\ g\ mol^{-1}}$$

From the reaction equation, we see that the ratio amount of CO:amount of Fe is 3:2, which is equivalent to the equation

$$\frac{amount\ of\ CO}{amount\ of\ Fe} = \frac{3}{2}$$

We can rearrange this to give

$$amount\ of\ Fe = \left(\frac{2}{3}\right) \times amount\ of\ CO$$

Substituting in our expression for the amounts of each of these gives

$$\frac{\text{mass of Fe}}{55.85\,\text{g mol}^{-1}} = \left(\frac{2}{3}\right) \times \frac{503\,\text{g}}{28.01\,\text{g mol}^{-1}}$$

This equation can be rearranged by multiplying both sides by 55.85 g mol^{-1} to give the final expression

$$\text{mass of Fe} = 55.85\,\text{g mol}^{-1} \times \left(\frac{2}{3}\right) \times \frac{503\,\text{g}}{28.011\,\text{g mol}^{-1}}$$

Evaluating this on a calculator gives the value 668.607 57 g. We can ignore the integers 2 and 3 in considering the appropriate number of significant figures, and are then left with 4, 3, and 4 figures, respectively, in the remaining quantities. It is therefore appropriate to give the answer to 3 significant figures, so our final answer should be 669 g, as we need to round up the final digit as the first digit deleted is 6.

It is worth noting that the calculator was only used at the very end of this process.

Note that if we had converted 2/3 to a decimal value earlier in the process, we might have been unable to determine the correct number of significant figures in the final answer.

Worked example 2.4

What volume of sulfuric acid of concentration 0.0502 mol dm^{-3} is required to neutralize 25.00 cm^3 of 0.0995 mol dm^{-3} sodium hydroxide?

Chemical background

Sulfuric acid is known as a dibasic acid because it contains two hydrogen atoms that may dissociate from the molecule to give the hydrogen ions, which confer acidity. In hydrochloric acid, there is only one such hydrogen, and the acid is known as a monobasic acid.

Sulfuric acid is one of the chemicals produced on a large industrial scale. It is used in making fertilizers and as a catalyst in the refining of petroleum.

Solution to worked example 2.4

The equation for the neutralization reaction taking place is

$$2\text{NaOH} + \text{H}_2\text{SO}_4 \rightarrow \text{Na}_2\text{SO}_4 + 2\text{H}_2\text{O}$$

So, we see that only 1 mol of sulfuric acid is required for every 1 mol of sodium hydroxide. Because the ratio amount of sodium hydroxide: amount of sulfuric acid is 2:1, we can write

$$\frac{\text{amount of sodium hydroxide}}{\text{amount of sulfuric acid}} = \frac{2}{1}$$

and then substitute the values into this equation to give

$$\frac{25.00\,\text{cm}^3 \times 0.0995\,\text{mol dm}^{-3}}{\text{volume of sulfuric acid} \times 0.0502\,\text{mol dm}^{-3}} = \frac{2}{1}$$

This rearranges to give the equation

$$\text{volume of sulfuric acid} = \frac{25.00\,\text{cm}^3 \times 0.0995\,\text{mol dm}^{-3}}{0.0502\,\text{mol dm}^{-3} \times 2}$$

which, on using a calculator, yields a volume of 24.775 896 41 cm^3. To determine the number of significant figures in this answer, we have to realize that the volume of sodium hydroxide will have been determined by a pipette and so the trailing zeros given will be significant. Ignoring the integer value of 2, we are left with combining 25.00 cm^3, which has at least 4 significant figures; 0.0995 mol dm^{-3}, which has 3 significant figures (ignore leading zeros); and 0.0502 mol dm^{-3}, which has 3 significant figures (ignore the leading zero but include the one "inside" the number), so our answer should be given to 3 significant figures. Rounding up the third of these leads to a final value for the volume of 24.8 cm^3.

2.4   Uncertainty in measurement

The rate constant for the hydrolysis of methyl acetate is 1.32×10^{-4} s at 25°C; the dissociation constant of hydrofluoric acid is 5.62×10^{-4} ; the enthalpy of formation of methane from its elements is –74.9 kJ mol^{-1}.

So far, we have addressed the question of how many figures are justified when quoting the numerical value of a particular quantity. In other words, we have seen how to specify the appropriate precision of a physical quantity. We now need to consider how we can judge the accuracy of our value.

The object of many physical chemistry experiments is to obtain the value of some numerical quantity. Examples are the rate constant for a reaction, the dissociation constant of an acid, or the enthalpy change accompanying a reaction. Frequently, in student work, the accepted values for these quantities are well documented, and a student can assess the uncertainty in his or her determination of a particular value by consulting the literature.

In genuine research work, this is not the case. You might be studying a new reaction or using a new technique and have no idea what the exact value is going to be. In this case, it is vital to be able to estimate the uncertainty associated with such a value. Even if the value is known, you need to be able to state whether your experimental value agrees with the literature value within the bounds of experimental uncertainty.

There are a number of approaches to the determination of experimental uncertainties. When carrying out research, the determination of a physical quantity would be repeated many times, and statistical methods can then be used as described later in this section. You are more likely to be in the position of having a fixed number of hours to perform an experiment, after which the demands of the course dictate that you move on to new work. The results of the class may be pooled to allow a statistical treatment but, more frequently, you will be left with a single determination of your value on which you need to estimate the uncertainty. In this case, you will have to estimate the probable uncertainty; we will see shortly that there are two ways in which this may be done.

2.4.1   Types of uncertainty

You may decide that you can read a burette to ± 0.05 cm^3, a balance to ± 0.0001 g, or a digital voltmeter to ± 0.001 V, for example. We use the $\pm$ symbol in the specification of uncertainties to show that they have an equal chance of being too high or too low by this amount.

The Greek letter Δ (capital delta) is conventionally used to denote a difference in two values of a quantity. It is much used in physical chemistry, and we will see it again in Chapter 3.

You need to be able to distinguish between three definitions of uncertainties. To illustrate these, suppose you have estimated that the uncertainty on a quantity X is ΔX. We can then define

- The absolute uncertainty as ΔX
- The fractional uncertainty as $\Delta X / X$
- The percentage uncertainty as $100 \times \Delta X / X$

Worked example 2.5

An actual volume of 25.00 cm^3 is measured as 25.15 cm^3. What are (a) the absolute uncertainty, (b) the fractional uncertainty, and (c) the percentage uncertainty in this quantity?

Chemical background

The uncertainty involved in this example is rather more than you should hope to achieve in practice! It helps to ensure that the burette is mounted vertically and that the effects of parallax are avoided when readings are taken, as shown in Figure 2.3.

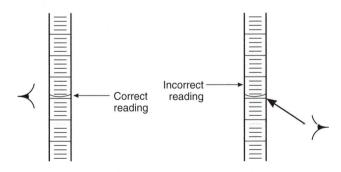

Figure 2.3   Effect of parallax when reading a burette.

Solution to worked example 2.5

(a) The absolute uncertainty is ±0.15 cm^3; we use the ± sign because we would not generally know in which direction the uncertainty was.

(b) The fractional uncertainty is

$$\pm \frac{0.15\,\text{cm}^3}{25.00\,\text{cm}^3} = \pm 0.006$$

(c) The percentage uncertainty will be 100 times the fractional uncertainty, which is ± 0.6%.

2.4.2   Combining Uncertainties

The determination of the value of a physical quantity in an experiment will usually involve several measurements, followed by one or more calculations involving these quantities. As these are combined using mathematical operations such as addition, subtraction, multiplication, and division, the corresponding uncertainties are also combined using a set of rules. Let us suppose that we have measured two quantities X and Y, and have estimated the uncertainties on them to be ΔX and ΔY respectively.

2.4.2.1   Determining the Maximum Possible Uncertainty

This can be thought of as the "common sense" approach. We do not need to remember any formulae, but we do need to be alert at every stage of a calculation.

(a) If X and Y are added, we have

$$\text{Maximum value of } (X + Y) = (X + \Delta X) + (Y + \Delta Y)$$

$$\text{Minimum value of } (X + Y) = (X - \Delta X) + (Y - \Delta Y)$$

(b) If Y is subtracted from X, we have

$$\text{Maximum value of } (X - Y) = (X + \Delta X) - (Y - \Delta Y)$$

$$\text{Minimum value of } (X - Y) = (X - \Delta X) - (Y + \Delta Y)$$

(c) If X and Y are multiplied, we have

$$\text{Maximum value of } XY = (X + \Delta X) \times (Y + \Delta Y)$$

$$\text{Minimum value of } XY = (X - \Delta X) \times (Y - \Delta Y)$$

(d) If X is divided by Y, we have

$$\text{Maximum value of } \frac{X}{Y} = \frac{X + \Delta X}{Y - \Delta Y}$$

$$\text{Minimum value of } \frac{X}{Y} = \frac{X - \Delta X}{Y + \Delta Y}$$

In each case the uncertainty is given as

0.5 (maximum value – minimum value).

2.4.2.2   Determining the Maximum Probable Uncertainty

(a) If X and Y are added, or one is subtracted from the other, the uncertainty on the result will be

$$\sqrt{\left(\Delta X\right)^2 + \left(\Delta Y\right)^2}$$

(b) If X and Y are multiplied, or one is divided by the other, the fractional uncertainty on the result will be

Use the x^2 calculator key to square each entity,

$$\sqrt{\left(\frac{\Delta X}{X}\right)^2 + \left(\frac{\Delta Y}{Y}\right)^2}$$

then the $\sqrt{\ }$ key to take the root of the resulting sum.

Following on from this, if X is raised to the power n, the fractional uncertainty on the product will be

$$\frac{\Delta X}{X}\sqrt{n}$$

Worked example 2.6

The initial and final burette readings in a titration are (5.00 ± 0.05) cm^3 and (31.85 ± 0.05) cm^3, respectively. What is the uncertainty in the titre?

Chemical background

Although it is probably more usual to refill a burette to the zero mark before starting each titration, there is an advantage to not doing this. Using a greater length of the burette will help to average out variations in the bore. It is important, however, to ensure that you do not run out of solution before the end point has been reached!

The capacity tolerances of precision volumetric glassware are specified by the National Physical Laboratory as "Grade A" or "Grade B." It should be possible to achieve an accuracy of within 0.05% with a pipette, and within 0.1% with a burette.

Solution to worked example 2.6

The value of the titre is found by subtracting the initial volume from the final volume, so we have

$$\text{Titre} = (31.85 - 5.00) \text{ cm}^3 = 26.85 \text{ cm}^3$$

We begin by calculating the maximum possible uncertainty. We have

$$\text{Maximum titre} = (31.90 - 4.95) \text{ cm}^3 = 26.95 \text{ cm}^3$$
$$\text{Minimum titre} = (31.80 - 5.05) \text{ cm}^3 = 26.75 \text{ cm}^3$$

with an uncertainty of

$$0.5 \times (26.95 - 26.75) \text{ cm}^3 = 0.5 \times 0.20 \text{ cm}^3 = 0.10 \text{ cm}^3$$

We calculate the maximum probable uncertainty as

$$\sqrt{(\Delta X)^2 + (\Delta Y)^2} = \sqrt{(0.05 \text{ cm}^3)^2 + (0.05 \text{ cm}^3)^2} = 0.07 \text{ cm}^3$$

Note that the maximum probable uncertainty is less than the maximum possible uncertainty. This will generally be the case; the formula for the maximum probable uncertainty takes account of the fact that an uncertainty will not normally have its maximum value. From this point we will calculate only the maximum probable uncertainty.

It is worth noting that the uncertainty value of 0.07 cm^3 has been obtained independently of the titre value and so is valid for any titration in which the burette has been consistently read to ±0 05 cm^3. We would express the final titre as (26.85 ± 0.07) cm^3. The brackets are used here to show that the units apply to both the titre value and its uncertainty but, by convention, they may be omitted, in which case the titre would be given as 26.85 ± 0.07 cm^3.

Worked example 2.7

Calculate the solubility of magnesium ethanedioate from the following information. A volume 14.85 ± 0.07 cm^3 of acidified potassium permanganate solution of concentration 0.0050 ± 0.000 05 mol dm^{-3} is required to oxidize 20.0 ± 0.1 cm^3 of a saturated solution of the oxalate. This reaction can be represented by the equation

$$5 \text{ C}_2\text{O}_4^{2-} + 2 \text{ MnO}_4^- + 16\text{H}^+ \rightarrow 10\text{CO}_2 + 8 \text{ H}_2\text{O} + 2\text{Mn}^{2+}$$

Chemical background

This is an example of a redox titration in which electron transfer is involved. The permanganate or manganate (VII) ion MnO_4^- is reduced to manganese (II) ions

$$\text{MnO}_4^- + 8 \text{ H}^+ + 5\text{e}^- \rightarrow \text{Mn}^{2+} + 4 \text{ H}_2\text{O}$$

whereas the ethanedioate ion $C_2O_4{}^{2-}$ is oxidized to carbon dioxide

$$C_2O_4{}^{2-} \rightarrow 2CO_2 + 2e^-$$

Solution to worked example 2.7

The only information that needs to be extracted from this equation is that the ratio amount of ethanedioate:amount of permanganate is 5:2. This gives us the equation

$$\frac{\text{amount of ethanedioate}}{\text{amount of permanganate}} = \frac{5}{2}$$

Substituting values for the concentration and volume leads to

$$\frac{20.0\,\text{cm}^3 \times \left[C_2O_4{}^{2-} \right]}{14.85\,\text{cm}^3 \times 0.050\,\text{mol dm}^{-3}} = \frac{5}{2}$$

which can be rearranged to give

$$\left[C_2O_4{}^{2-} \right] = \left(\frac{5}{2} \right) \times \frac{14.85\,\text{cm}^3 \times 0.0050\,\text{mol dm}^{-3}}{20.0\,\text{cm}^3}$$

The use of a calculator gives the value 9.3×10^{-3} mol dm^{-3}, applying the rules we have learned for significant figures and realizing that the trailing zero in the value of the permanganate concentration is significant.

If we now look at the equation we used to evaluate $[C_2O_4{}^{2-}]$, we see that the only mathematical operations involved are multiplication and division. Therefore, we can evaluate the overall uncertainty by using the rule for fractional uncertainties given earlier. First, we need to calculate the fractional uncertainty on each quantity.

$$\text{Fractional uncertainty on titre} = \pm \frac{0.07\,\text{cm}^3}{14.85\,\text{cm}^3} = \pm 0.0047$$

Fractional uncertainty on permanganate concentration

$$= \pm \frac{0.00005\,\text{mol dm}^{-3}}{0.0050\,\text{mol dm}^{-3}} = \pm 0.01$$

$$\text{Fractional uncertainty on ethanedioate volume} = \pm \frac{0.1\,\text{cm}^3}{20.0\,\text{cm}^3} = \pm 0.005$$

The stoichiometric ratio, $5/2$, is an exact number, so its associated uncertainty is zero. The overall fractional uncertainty is then

$$\sqrt{0.0047^2 + 0.0100^2 + 0.005^2} = \sqrt{0.000147} = 0.0121$$

We multiply this by our concentration value to give the absolute uncertainty

$$0.0121 \times 9.3 \times 10^{-3} \text{ mol dm}^{-3} = 0.11 \times 10^{-3} \text{ mol dm}^{-3}$$

The final value for the concentration is then given as $(9.3 \pm 0.1) \times 10^{-3}$ mol dm^{-3}. When scientific notation is involved, as here, it is usually clearer to quote the uncertainty as a number to the same power of 10 as the value to which it refers, even if it means that the uncertainty is not given as a number between 1 and 10.

Worked example 2.8

The concentrations of hydrogen, iodine, and hydrogen iodide in the gaseous equilibrium

$$H_2 + I_2 = 2HI$$

are related by the equilibrium constant K, so that if all other quantities are known, the hydrogen concentration can be determined from the equation

$$[H_2] = \frac{[HI]^2}{K[I_2]}$$

More data on this equilibrium reaction will be found in *Journal of the American Chemical Society*, **63**, 1377, 1941.

If the equilibrium constant K has been determined as 46.0 ± 0.2, calculate the concentration of hydrogen, with its associated uncertainty, when $[HI] = (17.2 \pm 0.2) \times 10^{-3}$ mol dm^{-3}, and $[I_2] = (2.9 \pm 0.1) \times 10^{-3}$ mol dm^{-3}.

Chemical background

This is an example of a chemical equilibrium that has been studied extensively. Although we have dealt with this in terms of concentrations, it is also possible to do so in terms of the partial pressures of each component. No matter what the starting concentrations or pressures, at equilibrium the relationship given will always be obeyed at a given temperature; at a different temperature, we would have to use a different value for the equilibrium constant K. One consequence of this relationship is that it does not matter whether we start with a mixture of hydrogen or iodine, or by allowing hydrogen iodide to decompose. The same equilibrium will be reached in each case.

One practical way in which this equilibrium could be studied would be to heat hydrogen iodide in a bulb of known volume as shown in Figure 2.4.

Figure 2.4   Measuring the equilibrium constant for the dissociation of hydrogen iodide.

Once equilibrium has been reached, this could then be cooled rapidly to room temperature to stop or "quench" the reaction. The concentration of iodine can then be determined by titration with sodium thiosulfate.

Solution to worked example 2.8

Using the equation given to evaluate the concentration of hydrogen gives us

$$[H_2] = \frac{(17.2 \times 10^{-3} \, \text{mol dm}^{-3})^2}{46.0 \times 2.9 \times 10^{-3} \, \text{mol dm}^{-3}}$$

which can be evaluated using a calculator to give the value of 2.218×10^{-3} mol dm^{-3}. Because we are going to evaluate an uncertainty limit from the information given, there is no point at this stage in worrying too much about the number of significant figures to quote. It is best to err on the side of caution, however, to make sure that we include enough significant figures to allow rounding at a later stage.

   Notice that, because only multiplication and division are involved in the equation given, we need to combine the fractional uncertainty on each quantity. As the equation contains $[HI]^2$ rather than $[HI]$, the first thing we will do is to evaluate the fractional uncertainty on $[HI]$, which is

$$\frac{0.2}{17.2} = 1.2 \times 10^{-2}$$

Because the concentrations and their associated uncertainties are multiplied by the same powers of 10, these can be ignored when we are dealing with fractional uncertainties. Using the equation above, we have

$$\text{Fractional uncertainty on } [HI]^2 = \sqrt{2} \times \text{fractional uncertainty on } [HI]$$

$$= \sqrt{2} \times 1.2 \times 10^{-2}$$

$$= 1.7 \times 10^{-2}$$

The remaining fractional uncertainties are straightforward to calculate:

$$\text{Fractional uncertainty on } K = \frac{0.2}{46.0} = 4.3 \times 10^{-3}$$

$$\text{Fractional uncertainty on } [I_2] = \frac{0.1}{2.9} = 3.4 \times 10^{-2}$$

We now combine these three fractional uncertainties using the rule for products and quotients:

$$\text{Fractional uncertainty on } [H_2] =$$

$$\sqrt{(1.7 \times 10^{-2})^2 + (4.3 \times 10^{-3})^2 + (3.4 \times 10^{-2})^2} = 0.0382$$

To obtain the absolute uncertainty on H_2, we need to multiply this fractional uncertainty by the value of $[H_2]$ itself.

$$\text{Absolute uncertainty on } [H_2] = 0.0382 \times 2.218 \times 10^{-3} \text{ mol dm}^{-3}$$

$$= 0.084\,850\,89 \times 10^{-3} \text{ mol dm}^{-3}$$

$$= 8.49 \times 10^{-5} \text{ mol dm}^{-3}$$

We can now write our final expression for the hydrogen concentration as

$$[H_2] = (2.22 \pm 0.08) \times 10^{-3} \text{ mol dm}^{-3}$$

2.4.3   Statistical treatment of uncertainties

Statistics is a discipline in itself, so we will do no more than look at the basics here to enable us to assign reasonable uncertainty limits. As noted earlier, the statistical treatment is only of use when we have repeated measurements of the same quantity. At our current level, it is only important to understand a few basic definitions of statistical quantities and to be able to apply them to sets of data.

To illustrate these definitions, consider the small data set

1.2	1.4	1.2	1.5	1.3

which contains five numbers. One way of representing these is by using the notation x_i, where i is an integer so that x_1 is the first number (1.2), x_2 is the second number (1.4), and so on.

The *arithmetic mean* is obtained by adding all the values of the quantity together and then dividing by the number of measurements taken. It can be denoted as

$$\bar{x} = \sum \frac{x_i}{n}$$

where n is the number of numbers.

Do not be put off by the use of some strange-looking mathematical symbols here. This is a good place to introduce some terminology as the concepts are still simple, and it will make life easier with some of the other quantities. The symbol Σ (a capital Greek sigma) simply means "add up." So, if we add the five numbers given above, we get 6.6, which is equal to Σx_i. We then divide this by n, the number of values which, in this case, is 5. This gives us a value for the arithmetic mean of

$$\frac{6.6}{5} = 1.3$$

using our previous rules for significant figures and noting that the integer 5 is taken as an exact number.

The *median* is obtained by arranging all the values obtained in ascending (or descending) order; if there are an odd number of values, the median is the middle one; otherwise, it is the arithmetic mean of the two middle values. So, arranging our data set in ascending order (lowest first) gives us

1.2	1.2	1.3	1.4	1.5

As we have an odd slumber of values, the central one is the third, and so the value of the median is 1.3.

The *mode* is the most frequently occurring value. Looking at our example, only 1.2 appears more than once, so this is the mode in this case.

The arithmetic mean, median, and mode can all be thought of as different kinds of "average." What none of these quantities tells us, however, is how the values are spread about these averages. To do this, we need to consider variance and standard deviation.

The variance σ^2 is the average of the squares of the deviations from the true value of the quantity being measured. However, because we are taking our estimate of this value as arithmetic mean, we need to divide by $n - 1$ instead of n, where n is the number of values we have. We could only divide by n if we were sure that our arithmetic mean was equal to the true mean; normally, it will only be an estimate. Using the terminology introduced above,

$$\sigma^2 = \sum \frac{(x_i - \bar{x})^2}{n - 1}$$

This means that we take each value of x_i in turn, subtract the arithmetic mean, square the result, and divide by $n - 1$. We then add each of these terms, the total sum being the variance. Its square root is the *standard deviation* σ, which is a measure of the spread of values about the mean. Notice that the standard deviation is of more use than the variance because it has the same units as the quantity being measured. We would expect 95% of all values to lie within two standard deviations of the mean so that two standard deviations can be taken as a reasonable uncertainty estimate for each individual reading.

We still need to address the problem of how closely our mean value is a measure of the true value. The best estimate would be obtained by taking several samples of size n; the standard error is then defined as

$$\frac{\sigma}{\sqrt{n}}$$

where σ has been determined as above. This expression is obtained by considering the variance of the sample mean, which is σ^2/n, the square root of this is taken because it has the same units as the measured quantity.

Doubling the standard error gives us a *confidence limit* of 95% on the mean; in other words, if we performed many determinations of the mean, we would expect 95% of the results to lie within this limit.

From the equations given for the standard deviation and the standard error of the mean, you should be able to see that the mean will have a lower estimated uncertainty than each individual measurement, as we might expect.

Worked example 2.9

Ten determinations of the dipole moment of hydrogen chloride gas gave the following values (in units of D):

1.048	1.047	1.053	1.048	1.051
1.053	1.045	1.051	1.047	1.047

Calculate (a) the arithmetic mean, (b) the median, (c) the mode, (d) the variance, and (e) the standard deviation of these values.

Chemical background

When placed in an electric field, all molecules have an induced dipole moment because the centres of gravity of positive and negative charges do not coincide. Polar molecules also have a permanent dipole moment (Figure 2.5), which can be determined from measurements of the dielectric constant, also called the relative permittivity. In the case of hydrogen chloride, this can be done using a specially constructed gas capacitance cell.

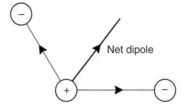

Figure 2.5   Net dipole resulting from two polar bonds.

Solution to worked example 2.9

(a) The sum of the values is 10.49 D, so the arithmetic mean is

$$\frac{10.49\,D}{10} = 1.049\,D$$

Notice that, by using the rules of significant figures (with 10 being an exact integer), we keep the same number of decimal places as in the original data.

(b) Rearranging the data in ascending order gives

1.045	1.047	1.047	1.047	1.048
1.048	1.051	1.051	1.053	1.053

Because there are an even number of data, we need to look at the fifth and sixth values; these are actually the same, so their arithmetic mean and the median of the data is 1.048 D.

(c) By inspection, the value of 1.047 D appears three times, and so this is the mode.

(d) To calculate the variance, we need to calculate the square of the deviation from the mean for each value that occurs:

$$\sigma^2 = \left(\frac{1}{10-1}\right) \times \begin{bmatrix} 1 \times (1.045-1.049)^2 \\ +3 \times (1.047-1.049)^2 \\ +2 \times (1.048-1.049)^2 \\ +2 \times (1.051-1.049)^2 \\ +2 \times (1.053-1.049)^2 \end{bmatrix} D^2$$

$$= \left(\frac{1}{9}\right)\left[1.6+1.2+0.2+0.8+3.2\right] \times 10^{-5}\,D^2$$

$$= 7.8 \times 10^{-6}\,D^2$$

(e) The standard deviation is the square root of this value, which is 0.003 D.

2.4.3.1 Statistics using a calculator

Although it is useful to have a knowledge of the underlying equations used, most of the simple statistics used in chemistry can be performed using the functions on a scientific calculator. However, as all types of calculator are slightly different, it is worth consulting your manual so that you know how to

The exact buttons that perform statistical calculations on a calculator vary from model to model. A typical sequence would be [MODE CLR] to select statistics mode, [M+] to enter each data item, then [SHIFT] followed by [2] to obtain the standard deviation.

1. Enter statistics mode
2. Enter data items
3. Obtain output

Worked example 2.10

The following determinations of the internuclear distance in hydrogen chloride were made (values in Å):

1.279	1.277	1.277	1.281
1.278	1.283	1.281	1.280

Use statistical methods to obtain an estimate of the H–Cl bond length with its uncertainty.

Chemical background

The H–Cl bond length can be determined by measuring the vibrational–rotational spectrum of hydrogen chloride gas, a part of which is shown in Figure 2.6. This gives the value of the rotational constant B, which is defined by

$$B = \frac{h}{8\pi^2 I}$$

where h is Planck's constant, and I is the moment of inertia defined by

$$I = \mu r^2$$

where μ is the reduced mass and r is internuclear distance.

The Ångström (Å) is a non-SI unit that still finds much favour among chemists working on structural problems. It is defined as 10^{-10} m, and consequently, a single bond between two carbon atoms has a length of 1.54 Å.

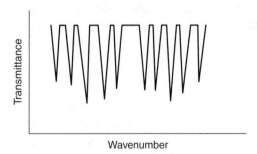

Figure 2.6   The vibration–rotation spectrum of hydrogen chloride gas.

The use of numbers around 1 and 2 seems to be preferred to the fractional nm (e.g., C–C length 0.154 nm) or larger pm (C–C length 154 pm).

Solution to worked example 2.10

We need to calculate the standard deviation of these data. The quickest way to do this is using a calculator, but here the manual working is also shown. We begin by calculating the arithmetic mean.

$$\text{Sum of data} = 10.236 \text{ Å}$$

$$\text{Arithmetic mean} = \frac{10.236}{8} = 1.280 \text{ Å}$$

The next stage is to obtain the variance.

$$\sigma^2 = \left(\frac{1}{8-1}\right) \times \Big[2 \times (1.277 - 1.280)^2 + 1 \times (1.278 - 1.280)^2 + 1$$

$$\times (1.279 - 1.280)^2 + 1 \times (1.280 - 1.280)^2 + 2 \times (1.281 - 1.280)^2$$

$$+ 1 \times (1.283 - 1.280)^2 \Big]$$

$$= \left(\frac{1}{7}\right) \times \Big[1.8 + 0.4 + 0.1 + 0.0 + 0.2 + 0.9 \Big] \times 10^{-5} \text{ Å}^2$$

$$= 4.86 \times 10^{-6} \text{ Å}^2$$

The standard deviation is the square root of this, 0.002 Å. It then follows that the standard error will be

$$\frac{\sigma}{\sqrt{n}} = \frac{0.002 \text{ Å}}{\sqrt{8}}$$

$$= \frac{0.002 \text{ Å}}{2.828}$$

$$= 7 \times 10^{-4} \text{ Å}$$

If we take twice the standard error as a reasonable uncertainty estimate, we have the H–Cl bond length as 1.280 ± 0.001 Å.

2.4.3.2   *Statistics using a spreadsheet*

Although the formula for the calculation of standard deviation is straightforward to apply, this can be rather tedious if we have a large number of data, even if a calculator is used. In such cases it makes sense to take advantage of one of the great features of computers, namely, that they are able to rapidly perform repetitive calculations. The most straightforward way to do this is to make use of the spreadsheet program found on most personal computers.

Worked example 2.11

The value for the C=O stretching frequency in a number of compounds belonging to several classes of carbonyl compounds is summarized in the table below in terms of wavenumber measured in cm^{-1}. Determine the standard deviation for this set of data.

Aldehyde	1722	1722	1739	1727	1728	1731	1727	1721	1738	1732
Ketone	1721	1708	1715	1713	1710	1719	1720	1710	1712	1708
Carboxylic acid	1721	1719	1716	1702	1717	1707	1711	1711	1719	1703
Ester	1742	1737	1749	1748	1739	1745	1749	1737	1736	1746
Carboxylate	1555	1601	1594	1592	1607	1559	1562	1568	1577	1580

Chemical background

Infrared spectroscopy is a very powerful technique for identifying compounds. This is because specified functional groups have characteristic vibrational frequencies in specified environments. It can be seen in the preceding table, in which the C=O stretching frequency is between 1720 and 1740 cm^{-1} in aldehydes, 1705 and 1725 cm^{-1} in ketones, 1700 and 1725 cm^{-1} in carboxylic acids, 1735 and 1750 cm^{-1} in esters, and 1550 and 1610 cm^{-1} in carboxylate ions.

Figure 2.7 Spreadsheet showing the input data for worked example 2.11.

Solution to worked example 2.11

The answer has been worked out using the spreadsheet Microsoft Excel, but the instructions for other spreadsheets will be similar. We begin by typing in the data, which leads to the spreadsheet shown in Figure 2.7. The text in cells A3 to A7 is not strictly required but is useful for tracking the data.

We now start to perform the stages in the calculation using the spreadsheet:

First, insert a blank column at the end of the original data. We will then begin our working in column M.

1. In cell M3, type =SUM(B3:K3). The = sign tells the program that we are about to input a formula. The SUM function adds the values in the range of cells specified in the following in brackets, in this case B3 to K3. Pressing Enter leads to the value 17 287. This is the sum of the values for the aldehydes.

2. We now copy this formula to the other compounds by dragging the corner of cell M3 down as far as M7.

3. Having obtained the sums for each class of compound, we now need to determine the overall sum. Label this by typing the text "Sum" in cell K9. Then, in M9 type =SUM(M3:M7).

4. Now type the text "Average" in cell K11. In M11, input =M9/50. Pressing Enter gives the value 1695.44.

5. We are now at the point where we need to calculate $(x_i - x)^2$ for each compound. In cell B16, input =(B3–M11)^2. This will take the value in cell B3 and subtract from it the value in cell M11, then square the result. Note that the latter is expressed as M11; this means that when the formula is copied, the value in M11 will always be used rather than the one in the cell relative to it, which is what we required in the case of B3, for example.

6. To copy the formula in cell B16, we copy the cell down as far as B20, then across to K20. This second table now contains the required value for each compound.

7. To add these values, type =SUM(B16:K16) in cell M16 to give the value 11 426.376. Then drag the corner of M16 down as far as M20, and in M22 type =SUM(M16:M20) to give 178 460.32.

8. We are now in a position to determine the variance, so we type an appropriate label in J25. Input the formula =M22/49 in cell M25 to give the value 3 642.047 3.

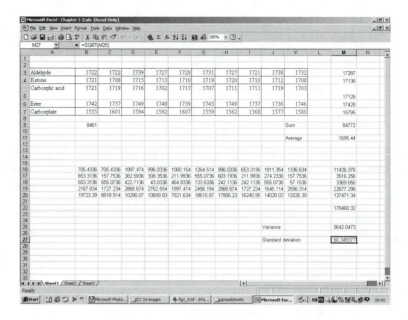

Figure 2.8   Spreadsheet showing the calculation for worked example 2.11.

9. We type a label for the standard deviation in J27, and in M27 the formula =SQRT(M25) to give 60.349 377.

The spreadsheet now appears as in Figure 2.8. We quote the standard deviation as 60 cm^{-1}.

Exercises

1. Write the following numbers to 2 significant figures:

   (a)   348   (b)   0.0012   (c)   0.5437   (d)   12.65

2. Write the following numbers to 2 decimal places:

   (a)   13.847   (b)   0.2549   (c)   99.543 21   (d)   0.007 54

3. Write the following numbers to 3 significant figures:

   (a)   78 521   (b)   0.006 7543   (c)   12.005 485   (d)   80 006

4. Write the following numbers to 1 decimal place:

   (a)   3.098   (b)   12.76   (c)   0.000 62   (d)   1.049

5. Give the results of the following calculations to the appropriate number of figures:

   (a) $12.36 + 3.2$ (b) $31.8 - 12.72$ (c) 1.9×1.235

(d) $\dfrac{4.673}{2.1}$ (e) $\dfrac{(34.2 - 1.75)}{2.34}$

6. Give the results of the following calculations to the appropriate number of significant figures:

(a) 2×32.74 (b) $3 \times 12.01 \times 32.1$ (c) $\left(\dfrac{5}{2}\right) \times \left(\dfrac{3.07}{6.18}\right)$

7. Give the results of the following calculations to the appropriate number of figures:

(a) $1.05 + 2.3 \times 4.08$ (b) $(2.49 - 1.06) \times 3.1$ (c) $4.989 - \dfrac{4.08}{1.249}$

8. Give the results of the following calculations to the appropriate number of figures:

(a) $9.714 - 2.62^2$ (b) $\dfrac{3.19 - 2.584}{1.2^3}$ (c) $\dfrac{(2.195 + 3.42)^2}{8.7148}$

9. A number is quoted as 32.6 ± 0.4. What are the (a) absolute uncertainty, (b) the fractional uncertainty, and (c) the percentage uncertainty?

10. The sides of a cuboid are measured as 21.0 ± 0.1 cm, 16.0 ± 0.1 cm, and 32.0 ± 0.2 cm. What are (a) the maximum possible absolute uncertainty, (b) the maximum possible fractional uncertainty, and (c) the maximum possible percentage uncertainty in the volume of the cuboid?

11. Repeat question 8, but this time calculate the maximum probable uncertainties in each case.

12. Three variables are measured as $x = 3.07 \pm 0.05$, $y = 4.29 \pm 0.08$, and $z = 2.37 \pm 0.10$. Determine the maximum probable uncertainty in

(a) $x + y - z$ (b) $xy + z$ (c) $x + \dfrac{y}{z}$

13. Four variables are measured as $a = 9.143 \pm 0.008$, $b = 6.413 \pm 0.007$, $c = 2.985 \pm 0.005$, and $d = 8.764 \pm 0.006$. Determine the maximum probable uncertainty in

(a) $a - b + \dfrac{c}{d}$ (b) $\dfrac{a}{b} + (d - c)^2$ (c) $\dfrac{(a - b)^3}{(c + d)^2}$

14. A car covered a distance of 9.2 ± 0.1 km in 295 ± 1 s. Use the formula

$$\text{speed} = \frac{\text{distance}}{\text{time}}$$

to calculate the average speed of the car and its associated maximum probable uncertainty.

15. Calculate (a) the arithmetic mean, (b) the median, and (c) the mode of the following numbers:

10.65	10.62	10.58	10.67	10.66	10.61	10.61

16. Determine the standard deviation of the following numbers by calculating each intermediate step:

23.72	24.01	23.86	24.09	23.68	23.99

17. Determine the standard error of the following numbers by calculating each intermediate step:

98.54	100.06	99.42	98.95	101.03	100.01	98.95	99.51

18. Use the statistical function on a calculator to determine the standard deviation of the following numbers:

3.79	3.61	3.84	3.76	3.72	3.82	3.66	3.73	3.65	3.75
3.83	3.71	3.62	3.65	3.66	3.83	3.62	3.81	3.85	3.74

19. Use the statistical function on a calculator to determine the standard deviation, and hence the standard error, of the following numbers:

72.1	74.8	73.7	74.1	73.2	74.3

20. Use a spreadsheet to determine the standard deviation of the following numbers:

19.45	19.42	19.51	19.55	19.53	19.45	19.45	19.56	19.52	19.49
19.44	19.50	19.50	19.56	19.47	19.44	19.49	19.48	19.44	19.47
19.46	19.43	19.49	19.51	19.47	19.49	19.55	19.54	19.50	19.49
19.55	19.53	19.47	19.48	19.52	19.55	19.53	19.47	19.45	19.44

Problems

1. Write the values of each of these quantities to (i) 3 significant figures and (ii) 2 decimal places:
   (a) a bond length of 1.542 Å

(b) a temperature of 25.01°C

(c) an enthalpy change of −432.876 kJ mol^{-1}

(d) the relative atomic mass of oxygen, 15.9994

(e) the gas constant, 8.314 J K^{-1} mol^{-1}

2. Give the results of the following calculations to the appropriate number of figures:

(a) $\text{density} = \dfrac{\text{mass}}{\text{volume}} = \dfrac{17.098\,\text{g}}{6.2\,\text{cm}^3}$

(b) mass of sample = 20.1452 g − 5.12 g

(c) internal energy $\Delta U = Q + W = 321.4\,\text{kJ mol}^{-1} - 162\,\text{kJ mol}^{-1}$

(d) $pV = nRT = 3.2\,\text{mol} \times 8.314\,\text{J K}^{-1}\,\text{mol}^{-1} \times 298\,\text{K}$

(e) equilibrium constant $K_c = \dfrac{0.106\,\text{mol dm}^{-3} \times 0.098\,\text{mol dm}^{-3}}{0.250\,\text{mol dm}^{-3}}$

3. Give the results of the following calculations to an appropriate number of figures:

(a) $\Delta G = -246\,\text{kJ mol}^{-1} - 298\,\text{K} \times 0.124\,\text{kJ K}^{-1}\,\text{mol}^{-1}$

(b) $C_p = 19.875\,\text{J K}^{-1}\,\text{mol}^{-1} + 5.021 \times 10^{-2}\,\text{J K}^{-2}\,\text{mol}^{-1} \times 305.4\,\text{K}$

(c) $\dfrac{1}{c} = \dfrac{1}{7.50 \times 10^{-3}\,\text{mol dm}^{-3}} + 0.059\,\text{dm}^3\,\text{mol}^{-1}\,\text{min}^{-1} \times 2.36\,\text{min}$

4. The virial equation can be expressed as

$$Z = 1 + \frac{Bp}{RT}$$

where p represents pressure, T the absolute temperature, and R the gas constant 8.314 J K^{-1} mol^{-1}. Calculate Z for helium for which $B = 11.8 \times 10^{-6}$ m^3 mol^{-1} at 298.15 K. Take the pressure as 250 kPa.

5. The overall entropy change for any process is equal to the sum of the entropy changes for the individual steps involved. In one experiment, these changes were measured as 3.90 ± 0.05 J K^{-1} mol^{-1}, 38.4 ± 0.1 J K^{-1} mol^{-1}, and 5.05 ± 0.05 J K^{-1} mol^{-1}. Determine the maximum possible uncertainty and the maximum probable uncertainty in the sum of these three entropy changes.

6. The volume V of an ideal gas can be found by rearranging the ideal gas equation, and is given by

$$V = \frac{nRT}{p}$$

where n is the amount of gas, R the gas constant (8.314 J K^{-1} mol^{-1}), T the absolute temperature, and p the pressure. Calculate the volume, with its associated uncertainty, for 2.35 ± 0.1 mol of gas at a temperature of 298.4 ± 0.1 K and a pressure of 1.033 ± 0.005 kPa.

7. Concentration may be expressed in terms of the mass percentage of solute, which is the percentage by mass of solute contained in a solution. A solution of sodium chloride was prepared by dissolving 2.67 ± 0.04 g of the salt in 84.3 ± 0.07 g of water. Determine the mass percentage of solute in this case, together with the maximum probable uncertainty.

8. Calculate the arithmetic mean, median, mode, variance, and standard deviation of the following values of the ionization energy of sodium (in kJ mol^{-1}):

495	497	493	493	496	492

9. Use a spreadsheet to determine the standard deviation of the following Cu–N bond lengths (in Å) in a series of compounds as determined by x-ray crystallography:

2.035	2.038	2.047	2.052	2.051	2.039	2.040	2.052	2.036	2.041
2.039	2.038	2.050	2.049	2.053	2.048	2.041	2.039	2.040	2.037
2.051	2.053	2.038	2.053	2.038	2.052	2.039	2.036	2.051	2.036

10. Calculate the standard error on the following measurements of the molar volume of an ideal gas (in dm^3):

22.406	22.412	22.410	22.408	22.409	22.414

Thermodynamics **3**

Thermodynamics is one of the branches of physical chemistry that students often find most confusing. It involves many different concepts, each of which seems to require a different mathematical approach and a different terminology. However, it is also a very important subject whose ideas underpin the whole of chemistry and without which a full appreciation of chemical concepts is not possible.

In this chapter, you will be taken through some of the main areas of thermodynamics and shown the mathematics required to deal with each of them. We will cover a number of different areas of mathematics, including possibly the first experience you have had of calculus. The terminology used can sometimes be off-putting, but each new symbol will be explained as we meet it. You will find the mathematics easier if you can remember the meanings of the symbols introduced.

3.1 Fractions and indices in the equilibrium constant

The equilibrium constant K is a fundamental quantity for any chemical reaction in which reactants and products are in equilibrium.

For the general chemical equilibrium

$$aA + bB = cC + dD$$

the equilibrium constant K_c is defined by

$$K_c = \frac{[C]^c [D]^d}{[A]^a [B]^b}$$

Equilibrium constants are important industrially because they allow the conditions that optimize production of the desired products to be determined.

In other words, we can write down an expression for the equilibrium constant as long as we known the equation for the equilibrium reaction.

Worked example 3.1

A volume 1.0 dm³ of a gaseous mixture contains 0.30 mol of CO, 0.10 mol of H_2, 0.020 mol of H_2O, and 0.059 mol of CH_4 at 1200 K. What is the equilibrium constant for

$$CO_{(g)} + 3H_{2(g)} = CH_{4(g)} + H_2O_{(g)}$$

at this temperature?

Chemical background

K_c actually specifies the equilibrium constant in terms of concentrations. A second equilibrium constant K_p is specified in terms of pressure. Although we might expect to calculate K_p for a gaseous reaction such as this, it is perfectly acceptable to calculate K_c if we know the amounts of each species present and the total volume.

Solution to worked example 3.1

We can immediately write down an expression for K_c, which is

$$K_c = \frac{[CH_4][H_2O]}{[CO][H_2]^3}$$

Because the total volume of the mixture is 1.0 dm^3, the amounts specified in the question in molar are numerically equal to their concentrations in mol dm^{-3}. The equilibrium constant is therefore

$$K_c = \frac{0.059\,\text{mol dm}^{-3} \times 0.020\,\text{mol dm}^{-3}}{0.30\,\text{mol dm}^{-3} \times (0.10\,\text{mol dm}^{-3})^3} = 3.9\,(\text{mol dm}^{-3})^{2-4}$$

$$= 3.9\,(\text{mol dm}^{-3})^{-2}$$

$$= 3.9\,\text{mol}^{-2}\,\text{dm}^6$$

$$= 3.9\,\text{dm}^6\,\text{mol}^{-2}$$

3.2   Bond enthalpies

The idea that each type of bond has an associated energy (more specifically, at constant pressure, enthalpy) gives us a straightforward way of estimating the enthalpy change for a given chemical reaction. If we assume that all chemical bonds in the reactants are broken, requiring positive enthalpy input, and that all chemical bonds in the products are formed, releasing negative enthalpy, then the overall enthalpy change will be the sum of the individual enthalpy changes.

3.2.1   Rearranging equations

The mathematical background to this was discussed in Chapter 1, Section 1.3, on page 7. If we have a relationship such as

$$w = x + y + z$$

and we wish to obtain an expression for y, we subtract x and z from each side to give

$$w - x - z = x + y + z - x - z$$

or

$$y = w - x - z$$

Worked example 3.2

The enthalpy change for the reaction

$$CH_{4(g)} + Cl_{2(g)} \rightarrow CH_3Cl_{(g)} + HCl_{(g)}$$

is -104 kJ mol^{-1}. If the bond enthalpy for C–H is 411 kJ mol^{-1}, for H–Cl is 428 kJ mol^{-1}, and for C–Cl is 327 kJ mol^{-1}, what is the bond enthalpy for Cl–Cl?

Chemical background

This is one in a series of reactions that add chlorine to methane to give successively CH_2Cl_2, $CHCl_3$, and CCl_4.

Bond enthalpies only allow an estimate of the overall enthalpy change for a reaction because they are average values measured across a large range of compounds.

Solution to worked example 3.2

In the reactants we need to break 4 C–H and 1 Cl–Cl bonds. The enthalpy required to do this is 4×411 kJ mol^{-1} + x, where x is the Cl–Cl bond enthalpy we wish to find. Evaluating the product gives 1644 kJ mol^{-1} + x.

In the products we form 3 C–H, 1 C–Cl, and 1 H–Cl bond. The total enthalpy released will thus be 3×411 kJ mol^{-1} + 327 kJ mol^{-1} + 428 kJ mol^{-1}, which is equal to 1988 kJ mol^{-1}; because this is enthalpy released, we take the negative value.

The overall enthalpy change for the reaction is then

$$1644 \text{ kJ mol}^{-1} + x - 1988 \text{ kJ mol}^{-1} = -104 \text{ kJ mol}^{-1}$$

To solve this equation for x we need to subtract 1644 kJ mol^{-1} and add 1988 kJ mol^{-1} to each side. This gives

$$1644 \text{ kJ mol}^{-1} + x - 1988 \text{ kJ mol}^{-1} - 1644 \text{ kJ mol}^{-1} + 1988 \text{ kJ mol}^{-1} =$$

$$-104 \text{ kJ mol}^{-1} - 1644 \text{ kJ mol}^{-1} + 1988 \text{ kJ mol}^{-1}$$

or

$$x = 240 \text{ kJ mol}^{-1}$$

3.3   The Born–Haber cycle

One of the fundamental ideas in thermodynamics is that the change in value of certain variables is equal to the sum of other smaller changes that give the same overall result. For example, when a chemical reaction involving some kind of molecular rearrangement takes place, the change in the enthalpy is the same as if we had dissociated all the reactants into atoms and then allowed them to recombine into the required products. This idea formed the basis of worked example 3.2 discussed earlier. Of course, we know that chemical reactions do not really happen in this way, but it is a useful concept that allows us to calculate quantities that are difficult to measure experimentally. An analogy from everyday life is that if we climb a mountain, the height we reach is independent of the path we choose to take.

Some of the thermodynamic variables for which this is true are those called the enthalpy, entropy, Gibbs free energy, and chemical potential; these are known as state functions. The two most common nonstate functions are work and heat. You will meet all of these variables at some point in your study of physical chemistry.

The mathematical consequence of using state functions is that we can calculate the value of a change in a particular fraction by adding together other changes that refer to a different route, even though the starting and end points are the same. Although this is very useful to chemists in enabling them to calculate enthalpy changes for reactions in which they cannot be measured, there is an important point to remember. We saw in the last chapter that any experimental measurement is subject to an uncertainty, and if we are adding together several experimental values, we are likely to obtain an overall uncertainty that is much higher. It follows that an enthalpy change calculated in this way is likely to be less certain than one that has been measured directly.

3.3.1   Combining uncertainties

The mathematical background to this has been discussed in Chapter 2, Section 2.4.2.2, on page 35; in most cases we will determine the maximum probable uncertainty. We saw that if two quantities are measured as $X + \Delta X$ and $Y + \Delta Y$, for sums and differences the absolute uncertainty is $\sqrt{(\Delta X)^2 + (\Delta Y)^2}$, whereas for products and quotients the fractional uncertainty is $\sqrt{\left(\dfrac{\Delta X}{X}\right)^2 + \left(\dfrac{\Delta Y}{Y}\right)^2}$.

The symbol MX is used to denote a generalized formula, where M represents a metal cation and X an anion. Examples would be CsCl and KF. These are 1:1 compounds in which there are equal numbers of cations and anions. A 1:2 structure would have the generalized formula MX_2, examples being CaF_2 and MnS_2.

Worked example 3.3

The standard lattice enthalpy $\Delta_{lattice}H^\circ$ of a 1 : 1 ionic crystal MX is given by the equation

$$\Delta_{lattice}H^\circ = -\Delta_f H^\circ + \Delta_{sub}H^\circ + \tfrac{1}{2}\Delta_{at}H^\circ + \Delta_{ion}H^\circ + \Delta_{ea}H^\circ$$

where $\Delta_f H°$ is the standard enthalpy of formation of MX, $\Delta_{sub} H°$ is the standard enthalpy of sublimation of M, $\Delta_{at} H°$ is the standard enthalpy of atomization of X_2, $\Delta_{ion} H°$ is the ionization enthalpy of M, and $\Delta_{ea} H°$ is the electron affinity of X.

Calculate the electron affinity of Cl, together with its associated uncertainty, using the following data:

Standard enthalpy of formation of NaCl$_{(s)}$	=	-411 ± 2 kJ mol^{-1}
Enthalpy of sublimation of Na	=	109 ± 1 kJ mol^{-1}
Enthalpy of atomization of Cl$_{2(g)}$	=	242 ± 1 kJ mol^{-1}
Ionization enthalpy of Na$_{(g)}$	=	494 ± 2 kJ mol^{-1}
Standard lattice enthalpy of NaCl	=	760 ± 3 kJ mol^{-1}

Chemical background

A consideration of the lattice enthalpy finds use when we are interested in the forces and energies within ionic crystals, as it is possible to compare the value with that obtained from theoretical calculations. In contrast to the problem here, we would normally calculate the value of the lattice enthalpy using experimental values of the other enthalpy changes; a direct experimental determination is not possible. The lattice enthalpy in this problem actually refers to the enthalpy change accompanying the reaction

$$NaCl_{(s)} \rightarrow Na^+_{(g)} + Cl^-_{(g)}$$

This contrasts with the enthalpy of formation, which is the enthalpy change for the reaction

$$Na_{(s)} + \tfrac{1}{2}Cl_{2(g)} \rightarrow NaCl_{(s)}$$

The other enthalpy changes required are the enthalpy of sublimation:

$$Na_{(s)} \rightarrow Na_{(g)}$$

and the enthalpy of ionization

$$Na_{(g)} \rightarrow Na^+_{(g)} + e^-$$

For anions, we can define the electron affinity, which in this case is the enthalpy change for the reaction

$$Cl^-_{(g)} \rightarrow Cl_{(g)} + e^-$$

This reaction is the reverse of what you might expect, and it is often written the other way round with accompanying phrases used to obtain the correct sign for the enthalpy change.

Sodium chloride exists as a regular cubic lattice, as shown in Figure 3.1.

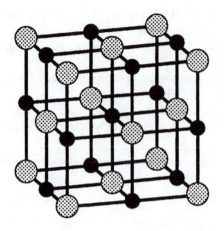

Figure 3.1 The crystal lattice of sodium chloride.

Solution to worked example 3.3

The calculation of $\Delta_{ea}H^\circ$ is straightforward and involves rearranging the given equation to leave this term as the subject. This gives

$$\Delta_{ea}H^\circ = \Delta_{lattice}H^\circ + \Delta_fH^\circ - \Delta_{sub}H^\circ - \tfrac{1}{2}\Delta_{at}H^\circ - \Delta_{ion}H^\circ$$

$$= [760 - 411 - 109 - (0.5 \times 242) - 494]\ \mathrm{kJ\ mol^{-1}}$$

$$= -375\ \mathrm{kJ\ mol^{-1}}$$

We can now turn our attention to determining the uncertainty in this quantity. First of all, let us look at the enthalpy of atomization of $Cl_{2(g)}$. This has been multiplied by 0.5, which we can take as being an exact constant with zero uncertainty. Therefore, the fractional uncertainty on $0.5\Delta_{at}H^\circ$ will be given by

$$\sqrt{\left(\frac{0}{0.5}\right)^2 + \left(\frac{1}{242}\right)^2} = \frac{1}{242}$$

which is the same as the fractional uncertainty on $\Delta_{at}H^\circ$. The absolute uncertainty in this quantity will therefore be

$$\left(\frac{1}{242}\right) \times 0.5 \times 242\ \mathrm{kJ\ mol^{-1}} = 0.5\ \mathrm{kJ\ mol^{-1}}$$

Use the calculator button to square each value, add the squares, and take the square root using ![√] .

Having done this, we have a set of uncertainties on five values that are combined by means of addition and subtraction. The absolute uncertainty is then given by

Thermodynamics   59

$$\sqrt{(3)^2 + (2)^2 + (1)^2 + (0.5)^2 + (2)^2}\ \text{kJ mol}^{-1} = \sqrt{18.25}\ \text{kJ mol}^{-1}$$

$$= 4.3\ \text{kJ mol}^{-1}$$

The value of $\Delta_{ea}H^\circ$ is now -375 ± 4 kJ mol^{-1}.

Note that the overall uncertainty is, inevitably, greater than any of the individual uncertainties used in the calculation. For this reason, a direct experimental measurement may be preferable to a calculated value such as this.

3.4   Heat capacity

The branch of thermodynamics concerned with heat changes is thermochemistry. The principal property that governs the thermal behaviour of a substance is the heat capacity, which can be defined at a simple level as the heat required to raise the temperature of the substance by 1 K. More precisely, we can identify two values of the heat capacity according to the conditions under which any change is brought about. They are defined as

$$C_p = \text{heat capacity at constant pressure}$$

and

$$C_v = \text{heat capacity at constant volume}$$

Frequently, we are concerned with the molar heat capacity, which then allows us to take account of the amount of substance present.

3.4.1   Expansion of brackets

Mathematical expressions can be made more concise by the use of brackets. This is generally useful for giving general expressions, but there are times when it is desirable to expand such an expression into its individual terms. For example, if we have an expression such as

$$a(b + c)$$

this is expanded by multiplying the quantity outside the bracket (in this case, a) with each term inside the bracket.

$$a(b + c) = ab + ac$$

If two brackets are multiplied together, such as

$$(a + b)(c + d)$$

then each term in the first bracket needs to be multiplied with each term in the second:

$$(a + b)(c + d) = a(c + d) + b(c + d)$$

Multiplying out again gives

$$(a + b)(c + d) = a(c + d) + b(c + d) = ac + ad + bc + bd$$

It may, of course, be possible to simplify the resulting expression in many cases.

Worked example 3.4

The value of the enthalpy change ΔH is given in terms of the heat capacity C_p and the temperature change ΔT as

$$\Delta H = C_p \Delta T$$

Write an expression for the absolute enthalpy H_2 at temperature T_2 in terms of the absolute enthalpy H_1 at temperature T_1 and the heat capacity.

Chemical background

This relationship arises directly from the definition of enthalpy, which is the quantity of heat supplied to a system at constant pressure. It can be calculated from the formula

$$H = U + pV$$

where U is the internal energy, p is the pressure, and V the volume.

The expression given in the problem is only valid when C_p can be considered to be independent of temperature. In the next worked example we will see an example of C_p depending on temperature.

Solution to worked example 3.4

We have already met the use of the Δ notation when discussing enthalpy changes in worked example 3.3. Although a quantity such as the enthalpy H does have absolute values, in thermodynamics we are frequently more interested in changes to these values. We use the notation ΔH to denote such an enthalpy change, where this is defined as

$$\Delta H = H_2 - H_1$$

The convention is always to subtract the initial value (in this case H_1) from the final value (H_2). Similarly, the change in temperature is given by

$$\Delta T = T_2 - T_1$$

and so the expression becomes

$$H_2 - H_1 = C_p(T_2 - T_1)$$

We now expand the right-hand side of this equation by multiplying the terms in the bracket by C_p to give

$$H_2 - H_1 = C_p T_2 - C_p T_1$$

We were asked to obtain an expression for H_2, so we rearrange the equation, making H_2 the subject, by adding the term H_1 to both sides.

$$H_2 = H_1 + C_p T_2 - C_p T_1$$

This is a simple problem, yet being able to perform manipulations like this is vital to a mastery of thermodynamics.

3.4.2   Polynomial expressions

You may have seen expressions such as

$$5x^3 + 2x^2 + 3x + 10$$

which is an example of a *polynomial expression in x*. The characteristic of a polynomial is that it contains a series of terms consisting of the variable (in this case x) raised to an integral power. The highest value of these powers is called the *order* of the expression, and the constants that multiply each of these terms are called *coefficients*. So, in the example above, the highest power is 3, so this expression is a polynomial of order 3. We can also note that

- The coefficient of x^3 is 5.
- The coefficient of x^2 is 2.
- The coefficient of x is 3.

The final value of 10 is called the *constant term*. Note that any of these terms can be missing in a particular example, so that the expression $2x^4 + 3x$ is also an example of a polynomial expression although it does not contain terms in x^3 or x^2, nor a constant.

Special names are given to polynomials with the lowest orders. These are:

- *Linear* for an expression of order 1
- *Quadratic* for an expression of order 2
- *Cubic* for an expression of order 3

Examples of graphical representations of each of these are shown in Figure 3.2.

Worked example 3.5

The heat capacities C_p for the species involved in the reaction

$$H_{2(g)} + \tfrac{1}{2}O_{2(g)} \rightarrow H_2O_{(g)}$$

can be expressed as a polynomial with the variable T representing absolute temperature as

$$C_p = a + bT + cT^2$$

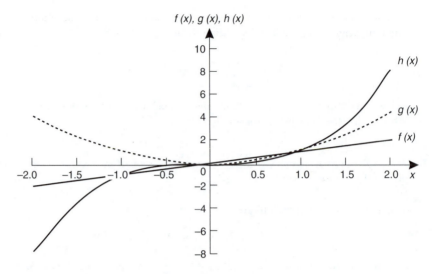

Figure 3.2   Graphs of representative linear, quadratic, and cubic functions, $f(x)$, $g(x)$, and $h(x)$, respectively.

where a, b, and c are constants. Use the values in the following table to obtain an expression for the change in heat capacity in terms of temperature when this reaction takes place.

	a	b	c
	$JK^{-1}\,mol^{-1}$	$10^3\,JK^{-2}\,mol^{-1}$	$10^7\,JK^{-3}\,mol^{-1}$
$H_{2(G)}$	29.07	−0.836	20.11
$O_{2(g)}$	25.72	12.98	−38.6
$H_2O_{(g)}$	30.36	9.61	11.8

Chemical background

The calculation of the expression is an important intermediate step when determining how an enthalpy change for a given reaction varies with temperature. As well as the form of polynomial shown here, an expression such as

$$C_p = a + bT + \frac{c}{T^2}$$

The chemical industry uses large quantities of water, the majority for removing heat. Pure water may be required for some chemical uses, and this can be produced by distillation.

can be used. Note that this is not actually a polynomial because of the final term, but it can be handled in exactly the same way as the expression in this problem.

Such expressions are able to fit experimental data to within 0.5% over a wide range of temperatures.

In fact, a mixture of hydrogen and oxygen gases will only react to give steam (gaseous water) if a spark is applied. This has the effect of supplying the activation energy for the reaction, which must always be considered in addition to arguments based on enthalpy changes.

Solution to worked example 3.5

This is the first problem we have met in which we have been required to extract data from a table. The form of the headings used here may seem rather strange but, in fact, this is the best way of presenting data. Let us look at the value of b for $O_{2(g)}$, the value in the very centre of the table. All we have to do is to equate the pure number in the table with its column heading to give

$$\frac{b}{10^3 \, J \, K^{-2} \, mol^{-1}} = 12.98$$

If we multiply each side of this equation by the units, we obtain

$$b = 12.98 \times 10^3 \, J \, K^{-2} \, mol^{-1}$$

and, if required, we can convert to correct scientific notation by rewriting this as a number between 1 and 10 and adjusting the power of 10 as

$$b = 1.298 \times 10^4 \, J \, K^{-2} \, mol^{-1}$$

A similar technique is used to read values from graphs, whose axes labels should be given in a similar fashion to these table headings.

The answer to this problem will consist of another polynomial expression whose terms represent the difference in heat capacities between the reactants and the products. It will be of the form

$$\Delta C_p = \Delta a + T\Delta b + T^2 \Delta c$$

where

$$\Delta C_p = \Sigma \, C_p \, (\text{products}) - \Sigma \, C_p \, (\text{reactants})$$

Notice that we are using notation (Δ and Σ) here to make the working more concise. Remember that Δ is used to represent a change, and that this is always taken as the final value minus the initial value. In this case, that is equivalent to taking products (final state) minus reactants (initial state). You may recall from Chapter 2, Section 2.4.3, on page 40 that the Σ symbol means "add up," so we are summing the contributions from the products and from the reactants for each of the coefficients a, b, and c. Therefore, we have

$$\Delta a = \Sigma\, a \text{ (products)} - \Sigma\, a \text{ (reactants)}$$

$$= 30.36 - [29.07 + (0.5 \times 25.72)] \text{ J K}^{-1} \text{ mol}^{-1}$$

$$= -11.57 \text{ J K}^{-1} \text{ mol}^{-1}$$

Note that the contribution from $O_{2(g)}$ is multiplied by 0.5 because this species appears in the reaction equation with a stoichiometry coefficient of ½. Also,

$$\Delta b = \Sigma\, b \text{ (products)} - \Sigma\, b \text{ (reactants)}$$
$$= (9.61 - [-0.836 + (0.5 \times 12.98)]) \times 10^3 \text{ J K}^{-2} \text{ mol}^{-1}$$
$$= 3.96 \times 10^3 \text{ J K}^{-2} \text{ mol}^{-1}$$

and

$$\Delta c = \Sigma\, c \text{ (products)} - \Sigma\, c \text{ (reactants)}$$
$$= (11.8 - [20.11 + (0.5 \times -38.6)]) \times 10^7 \text{ J K}^{-3} \text{ mol}^{-1}$$
$$= 11.0 \times 10^7 \text{ J K}^{-3} \text{ mol}^{-1}$$

It is slightly neater to write this using correct scientific notation as

$$\Delta c = 1.10 \times 10^8 \text{ J K}^{-3} \text{ mol}^{-1}$$

Now substituting into our equation for ΔC_p gives

$$\Delta C_p = -11.57 \text{ J K}^{-1} \text{ mol}^{-1} + (3.96 \times 10^3 \text{ J K}^{-2} \text{ mol}^{-1})\,T$$
$$+ (1.10 \times 10^8 \text{ J K}^{-3} \text{ mol}^{-1})T^2$$

Note the use of brackets to simplify this expression when the units are included. Another way of doing this would be to divide through by the common unit and write

$$\frac{\Delta C_p}{\text{J K}^{-1}\text{mol}^{-1}} = -11.57 + (3.96 \times 10^3 \text{ K}^{-1})T + (1.10 \times 10^8 \text{ K}^{-2})T^2$$

Personal preference dictates how you should leave this expression.

3.4.3   Functions

So far in this section we have seen that heat capacity can be regarded as a constant in some circumstances, whereas in others we can specify a dependence on temperature. In the latter case, we would say that heat capacity is a function of temperature. It can then be denoted as $C_p(T)$, where the T in brackets indicates that the value of T determines that of C_p. In general, we use the notation $f(x)$ to denote a function f of the variable x.

As an example, if $f(x)$ is defined as

$$f(x) = 8x^2 + 2x + 9$$

we can calculate the value of the function for any value of x. Using this notation, $f(3)$ would be the value of this function when $x = 3$, and would be given by substituting for x on both sides of the defining equation to give

$$f(3) = (8 \times 3^2) + (2 \times 3) + 9$$
$$= (8 \times 9) + (2 \times 3) + 9$$
$$= 72 + 6 + 9$$
$$= 87$$

Worked example 3.6

If the heat capacity C_p of solid lead is given by the expression

$$C_p (T) = 22.13 \text{ J K}^{-1} \text{ mol}^{-1} + (1.172 \times 10^{-2} \text{ J K}^{-2} \text{ mol}^{-1})T$$

$$+ \frac{(9.6 \times 10^4 \text{ J K mol}^{-1})}{T^2}$$

determine the heat capacity at 298 K.

Chemical background

Observation showed that different quantities of heat were required to raise the temperature of equal amounts of different substances to the same extent. More careful experiments then showed that this heat capacity actually varied with the temperature. The SI unit of molar heat capacity is J K^{-1} mol^{-1}; heat changes can be measured accurately by electrical means and $1 \text{ J} = 1 \text{ V} \times 1 \text{ C}$.

Solution to worked example 3.6

We actually need to calculate C_p (298 K), which is given by a straightforward substitution into both sides of the defining equation:

$$C_p(298 \text{ K}) = 22.13 \text{ J K}^{-1} \text{ mol}^{-1} + (1.172 \times 10^{-2} \text{ J K}^{-2} \text{ mol}^{-1} \times 298 \text{ K})$$
$$+ \frac{(9.6 \times 10^4 \text{ J K mol}^{-1})}{(298 \text{ K})^2}$$

$$= 22.13 \text{ J K}^{-1} \text{ mol}^{-1} + 3.492 \text{ J K}^{-1} \text{ mol}^{-1} + 1.081 \text{ J K}^{-1} \text{ mol}^{-1}$$

$$= (22.13 + 3.492 + 1.081) \text{ J K}^{-1} \text{ mol}^{-1}$$

$$= 26.70 \text{ J K}^{-1} \text{ mol}^{-1}$$

3.5   Clapeyron equation

There is a certain amount of vapour above every solid and liquid, and if this is not free to escape, as in a closed vessel, equilibrium will be reached so that as many molecules leave the surface as reenter. At this equilibrium, the pressure above the solid or liquid is known as the vapour pressure. A study of vapour pressure is important because it can provide us with information about the solid or liquid.

The Clapeyron equation applies to any phase transition of a pure substance and is expressed by the equation

$$\frac{dp}{dT} = \frac{\Delta_t H}{T \Delta_t V}$$

where $\Delta_t H$ and $\Delta_t V$ are the enthalpy and volume changes, respectively, which accompany a phase transition t at absolute temperature T. The left-hand side introduces a notation that we have not met before; it needs to be treated as a complete quantity, and the letter d, on its own, has no meaning in this equation. The quantity

$$\frac{dp}{dT}$$

The graph of pressure against temperature for a substance is known as its phase diagram and can be used to indicate where a solid, liquid, or gas will be found for given values of these variables. A typical one-component phase diagram is shown in Figure 3.3.

is called the *derivative of p with respect to T.* As we will see in the following text, this function is equal to the gradient of the tangent of the graph of p against T.

3.5.1   Differentiation

If we consider a function such as

$$f(x) = 3x^2 + 1$$

Figure 3.3   A typical phase diagram.

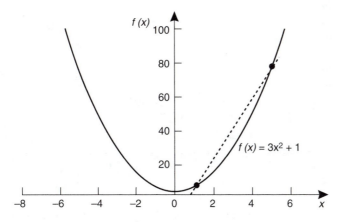

Figure 3.4   The graph of $f(x) = 3x^2 + 1$.

whose graph is shown in Figure 3.4, we can calculate an average rate of change between any two values of x, such as $x = 1$ and $x = 5$. This is found by dividing the difference between the values of the function at the two values of x by the difference in the two values of x:

$$\text{average rate of change} = \frac{f(5) - f(1)}{5 - 1}$$

because

$$f(1) = (3 \times 1^2) + 1 = 3 + 1 = 4$$

and

$$f(5) = (3 \times 5^2) + 1 = 75 + 1 = 76$$

we have

$$\text{average rate of change} = \frac{76 - 4}{5 - 1}$$

$$= \frac{72}{4}$$

$$= 18$$

This is equivalent to calculating the gradient of the line drawn between $f(1)$ and $f(5)$ on the graph shown in Figure 3.4.

It is also possible to determine the instantaneous rate of change of this function at a specified value of x, say $x = 2$. This is equivalent to determining the gradient of the tangent drawn to the curve at this value, as shown in Figure 3.5.

More generally, suppose that we wish to find the instantaneous rate of change, or gradient, when x has the specified value x_0. This is specified as point

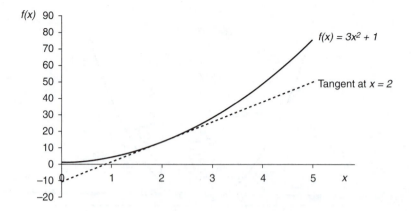

Figure 3.5   The gradient of the tangent to the function $f(x) = 3x^2 + 1$, at the point where $x = 2$.

Figure 3.6   Definition of the derivative $\dfrac{df(x)}{dx}$.

P in Figure 3.6. A second point Q on the curve defined by $f(x)$ has $x = x_0 + h$. The gradient of the line drawn between P and Q will therefore be

$$\frac{f(x_0 + h) - f(x_0)}{(x_0 + h) - x_0} = \frac{f(x_0 + h) - f(x_0)}{h}$$

Imagine now that Q slides down the curve until it is almost at the same place as P. This has the effect of making the value of h very small. In our minds, we can make h as small as we like as long as it does not become exactly zero. This is because if h were to become zero, the denominator would be zero, but division by zero is not defined. However, making h as close to zero as possible will give a very close approximation to the value of the gradient that we require. In mathematical terminology, we would say that we wish to calculate the above expression in the limit as h tends to zero. This is written as

$$\text{Gradient of tangent} = \lim_{h \to 0} \frac{f(x_0 + h) - f(x_0)}{h} \quad \text{when } x = x_0.$$

More usually, we would write this as

$$\frac{df(x)}{dx} = \lim_{h \to 0} \frac{f(x_0 + h) - f(x_0)}{h}$$

where

$$\frac{df(x)}{dx}$$

is the derivative of $f(x)$ with respect to x, in this case when $x = x_0$. This is spoken as "d f of x by d x." It is important to remember that this represents the gradient of a graph of $f(x)$ against x at $x = x_0$.

So far, the method of calculating a derivative involves taking a limit of a mathematical expression. In practice, this is not necessary as we can use a number of standard derivatives. In the case of a power of x, the derivative of x^n is nx^{n-1}, where n is a constant.

The following table shows this rule for $n = -2, -1, 0, 1, 2, 3, 4$.

n	Power of x	Derivative
-2	x^{-2}	$-2x^{-3}$
-1	x^{-1}	$-x^{-2}$
0	$x^0 = 1$	$0 \times x^{-1} = 0$
1	x^1	$1 \times x^0 = 1$
2	x^2	$2 \times x^1 = 2x$
3	x^3	$3 \times x^2 = 3x^2$
4	x^4	$4 \times x^3 = 4x^3$

When a power of x has a coefficient, e.g., ax^n, the derivative is anx^{n-1}, so that, for example

$$\frac{d(3x^4)}{dx} = 3(4x^{4-1}) = 12x^3$$

Notice that a constant on its own can be thought of as a term where the constant is multiplied by an x^0. The derivative of x^0 will always be zero, so that, for instance

$$\frac{d(6)}{dx} = \frac{d(6x^0)}{dx} = 6\frac{dx^0}{dx} = 6 \times 0 = 0$$

The rule given is readily extended to more complicated functions. If, for example, we have a function $f(x)$ defined as

$$f(x) = 2x^3 + 4x^2 + 7x + 2$$

we can apply the standard derivatives to each of these in turn. Notice the notation used for describing each of these derivatives:

$$\frac{d(2x^3)}{dx} = 2\frac{d(x^3)}{dx} = 2 \times 3x^2 = 6x^2$$

$$\frac{d(4x^2)}{dx} = 4\frac{d(x^2)}{dx} = 4 \times 2x = 8x$$

$$\frac{d(7x)}{dx} = 7\frac{d(x)}{dx} = 7 \times 1 = 7$$

$$\frac{d(2)}{dx} = 0$$

Putting this together gives the derivative as

$$\frac{df(x)}{dx} = 6x^2 + 8x + 7$$

Worked example 3.7

The triple point of iodine, I_2, occurs at a temperature of 114°C and a pressure of 12 kPa. If the enthalpies of fusion $\Delta_{fus}H$ and vaporization $\Delta_{vap}H$ are 15.52 kJ mol^{-1} and 41.80 kJ mol^{-1}, respectively, sketch the pressure–temperature graph in the region of the triple point. Take the densities of the solid and liquid to be 4.930 g cm^{-3} and 2.153 g cm^{-3}, respectively, and the molar mass of iodine atoms to be 126.9 g mol^{-1}. Assume that iodine vapour behaves as an ideal gas, and that the enthalpy of sublimation $\Delta_{sub}H$ is given by the equation:

$$\Delta_{sub}H = \Delta_{fus}H + \Delta_{vap}H$$

Chemical background

The term sublimation refers to a change directly from solid to vapour without passing through the liquid phase. Such behaviour can be predicted from the graph of pressure against temperature showing the equilibria between phases. From the equation given above, we can see that the enthalpy change will be the same when we pass from the solid to vapour phases, regardless of whether an intermediate liquid phase is involved.

Solid carbon dioxide also sublimes.

The phase diagram obtained in this question has the three lines representing phase boundaries, all with positive gradients. If we repeated the exercise for water, we would find that the slope of the line representing the solid-to-liquid transition would be negative. This is quite unusual and is a result of the reduction in the volume of ice on melting.

Solution to worked example 3.7

The question gives us all the information we need to evaluate the right-hand side of the Clapeyron equation

$$\frac{dp}{dT} = \frac{\Delta_t H}{T \Delta_t V}$$

which is equal to the left-hand side, i.e., the gradient of the graph of p and T. Because we are only asked about values close to the triple point, we can assume that the gradients we obtain are constant in this region of interest. We need to consider the three possible phase transitions in turn: solid to liquid; liquid to vapour; and solid to vapour.

Solid to liquid
The volume change ΔV will be

$$\Delta V = V_{\text{liquid}} - V_{\text{solid}}$$

where each volume can be obtained from the formula

$$V = \frac{m}{\rho}$$

with m representing mass and ρ density. Because 1 mol of I_2 has a mass of 2×126.9 g or 253.8 g, we have

$$V_{\text{liquid}} = \frac{253.8\,\text{g mol}^{-1}}{2.153\,\text{g cm}^{-3}} = 117.9\,\text{cm}^3\,\text{mol}^{-1}$$

$$V_{\text{solid}} = \frac{253.8\,\text{g mol}^{-1}}{4.930\,\text{g cm}^{-3}} = 51.48\,\text{cm}^3\,\text{mol}^{-1}$$

so that

$$\Delta V = (117.9 - 51.48)\,\text{cm}^3\,\text{mol}^{-1} = 66.4\,\text{cm}^3\,\text{mol}^{-1}$$

We also need to convert the given temperature value of 114°C to units of K.

$$\frac{T}{K} = \frac{114°C}{°C} + 273 = 387$$

so

$$T = 387\,\text{K}$$

Note that the more precise conversion factor (273.15 rather than 273) is not used because the temperature is only given to the nearest °C.

The Kelvin temperature scale does not use negative numbers as its zero is the lowest possible temperature that can be attained. Temperatures measured on this scale are known as absolute temperatures (symbol T) and are related to those on the Celsius scale (symbol θ) by the expression

$$\frac{T}{K} = \frac{\theta}{°C} + 273.15.$$

Substituting into the Clapeyron equation now gives

$$\frac{dp}{dT} = \frac{15.52\,\text{kJ}\,\text{mol}^{-1}}{387\,\text{K} \times 66.4\,\text{cm}^3\,\text{mol}^{-1}}$$

$$= \frac{15.52 \times 10^3\,\text{J}\,\text{mol}^{-1}}{387\,\text{K} \times 66.4 \times 10^{-6}\,\text{m}^3\,\text{mol}^{-1}}$$

$$= 6.04 \times 10^5\,\text{Pa}\,\text{K}^{-1}$$

because 1 kJ = 10^3 J, 1 cm^3 = 10^{-6} m^3, and 1 Pa = 1 N m^{-2} = 1 N m m^{-3} = 1 J m^{-3}.

Note the conversion from centimetres to metres:

$$100\ \text{cm} = 1\ \text{m}$$

$$1\ \text{cm} = 10^{-2}\ \text{m}$$

Cubing each side gives us

$$(1\ \text{cm})^3 = (10^{-2}\ \text{m})^3$$

or

$$1^3\ \text{cm}^3 = 10^{-6}\ \text{m}^3$$

Notice that $(10^{-2})^3 = 10^{-6}$. Now, $1^3 = 1$, so

$$1\ \text{cm}^3 = 10^{-6}\ \text{m}^3$$

Liquid to vapour

Because the question tells us to treat the iodine vapour as an ideal gas, we can use the ideal gas equation to calculate its volume at the triple point. If

$$pV = nRT$$

this can be rearranged (by dividing both sides by p) to give

$$V = \frac{nRT}{p}$$

Substituting values for 1 mol at the triple point gives us

$$V = \frac{1\,\text{mol} \times 8.314\,\text{J}\,\text{K}^{-1}\,\text{mol}^{-1} \times 387\,\text{K}}{12\,\text{kPa}}$$

$$= \frac{8.314 \times 387}{12} \frac{\text{J}}{\text{kPa}}$$

$$= 268.1265 \times \frac{\text{J}}{10^{3}\,\text{J}\,\text{m}^{-3}}$$

$$= 0.2681\,\text{m}^{3}$$

taking the value of the gas constant R as 8.314 J K^{-1} mol^{-1} and because 1 kPa = 10^3 Pa = 10^3 J m^{-3}. We are now able to calculate the volume change for the vaporization process for 1 mol:

$$\Delta V = V_{\text{vapour}} - V_{\text{liquid}}$$

$$= 0.2681\ \text{m}^3\ \text{mol}^{-1} - 117.9\ \text{cm}^3\ \text{mol}^{-1}$$

We need to convert these two values to the same units. Generally, it is preferable to work in base units (such as metres) rather than multiples (such as centimetres), so we choose to convert cm^3 to m^3, and use the rule 1 cm^3 = 10^{-6} m^3. Then

$$117.9\ \text{cm}^3 = 117.9 \times 10^{-6}\ \text{m}^3$$

$$= 1.179 \times 10^{-4}\ \text{m}^3$$

If we compare this with the vapour volume of 0.2681 m^3, we see that it is negligible and we are quite justified in approximating the volume change on vaporization to be equal to the volume of the vapour. In this case, we will take ΔV as 0.2681 m^3 mol^{-1}. Substituting into the Clapeyron equation leads us to

$$\frac{dp}{dT} = \frac{41.80\,\text{kJ}\,\text{mol}^{-1}}{387\,\text{K} \times 0.2681\,\text{m}^3\,\text{mol}^{-1}}$$

$$= \frac{41.80 \times 10^3\,\text{J}\,\text{mol}^{-1}}{387 \times 0.2681\,\text{K}\,\text{m}^3\,\text{mol}^{-1}}$$

$$= 402.9\,\text{J}\,\text{K}^{-1}\,\text{m}^{-3}$$

$$= 402.9\,\text{J}\,\text{m}^{-3}\,\text{K}^{-1}$$

$$= 402.9\,\text{N}\,\text{m}\,\text{m}^{-3}\,\text{K}^{-1}$$

$$= 402.9\,\text{N}\,\text{m}^{-2}\,\text{K}^{-1}$$

$$= 402.9\,\text{Pa}\,\text{K}^{-1}$$

Solid to vapour

Because the volume of the solid will be even less than that of the liquid, we are again justified in taking the volume change as being equal to the volume of the vapour, so we will set ΔV to 0.2681 m³ mol⁻¹. The enthalpy change $\Delta_{sub}H$ is given by the expression in the question, which evaluates to give

$$\Delta_{sub}H = \Delta_{fus}H + \Delta_{vap}H$$

$$= 15.52 \text{ kJ mol}^{-1} + 41.80 \text{ kJ mol}^{-1}$$

$$= 57.32 \text{ kJ mol}^{-1}$$

Substitution into the Clapeyron equation now gives

$$\frac{dp}{dT} = \frac{57.32 \text{ kJ mol}^{-1}}{387 \text{ K} \times 0.2681 \text{ m}^3 \text{ mol}^{-1}}$$

$$= \frac{57.32 \times 10^3 \text{ J mol}^{-1}}{387 \times 0.2681 \text{ m}^3 \text{ mol}^{-1}}$$

$$= 552.5 \text{ Pa K}^{-1}$$

Using these three gradient values, we now need to draw lines of appropriate slope that pass through the triple point. Notice that the value of dp/dT for the fusion process is much larger than the other two, so this can be represented by a vertical line. The value for vaporization is

$$\frac{402.9}{552.5} = 0.729$$

or 0.7 times the value for sublimation, which allows a simple qualitative representation to be drawn.

One possible graph, with values included, is shown in Figure 3.7.

Figure 3.7   Phase diagram for iodine.

3.6   Clausius–Clapeyron equation

The Clausius–Clapeyron equation relates the vapour pressures p_1 and p_2 at two temperatures T_1 and T_2 to the enthalpy of vaporization $\Delta_{vap}H$:

$$\ln\left(\frac{p_2}{p_1}\right) = \frac{\Delta_{vap}H}{R}\left[\left(\frac{1}{T_1}\right) - \left(\frac{1}{T_2}\right)\right]$$

Here we meet the ln symbol for the first time. This is an abbreviation for natural logarithm, and in this equation is telling us to "take the natural logarithm of the quantity p_2/p_1."

3.6.1   Logarithms

Historically, logarithms were used to make the processes of multiplication and division easier, but with the advent of electronic calculators, this use is now virtually redundant. However, their use is of much more fundamental importance than that, particularly when dealing with quantities that may span a very large range of values. There are two types of logarithms with which we need to be concerned: natural logarithms (ln), and logarithms to base 10 (log).

If three numbers a, b, and c are related such that

$$a = b^c$$

$$\log_b a = c$$

where b is called the base of the logarithm. Thus, if we choose b to have the value 10, then if

$$a = 10^c$$

it follows that

$$\log a = c$$

because we normally use the term "log" to denote logarithms to the base 10. On the other hand, if b is set to the number known as e, which has the value 2.718 28 …, we have

$$a = e^c$$

and

$$\ln a = c$$

because logarithms to the base "e" are called natural logarithms, and are denoted by the term "ln."

Natural logarithms frequently arise from the mathematical analysis of problems in physical chemistry, whereas logarithms to the base 10 are of more use when we want to display a large range of data. The two types of logarithms are related by the expression

$$\ln x = 2\ 303 \log x$$

We will meet an example of the natural logarithm emerging from the analysis of a physical chemistry problem in Chapter 5. This frequently arises from expressions involving a term in $1/x$, where x is a variable.

Worked example 3.8

The vapour pressure p of neon as a function of temperature θ is as follows:

$\dfrac{\theta}{°C}$	$\dfrac{p}{\text{mm Hg}}$
−228.7	19 800
−233.6	10 040
−240.2	3 170
−243.7	1 435
−245.7	816
−247.3	486
−248.5	325

By using an appropriate mathematical transformation, display this data graphically.

Chemical background

These data are a selection of those that were obtained when this study was performed. The complete set was fitted to an expression of the form

$$\log\left(\frac{p}{\text{mm Hg}}\right) = \frac{0.05223A}{T} + B + CT$$

to give the values of the constants

$$A = -1615.5 \text{ K}$$

$$B = 5.699\ 91$$

$$C = 0.011\ 1800 \text{ K}^{-1}$$

Notice the use of the heading $\log(p/\text{mm Hg})$ in the preceding equation. This is because it is only possible to take the logarithm of a pure number, i.e., one that does not have any units. If all the pressure values are divided by the units, we end up with pure numbers. For example, if p = 19 800 mm Hg, we can write

$$\frac{p}{\text{mm Hg}} = \frac{19800 \text{ mm Hg}}{\text{mm Hg}} = 19800$$

which is a similar technique to the one we used in worked example 3.5.

We evaluate logarithms to the base 10 using the

10^x

log button on a calculator.

Solution to worked example 3.8

We will consider the details of graph drawing shortly but for the moment merely note that pressure is plotted on the vertical y-axis and temperature

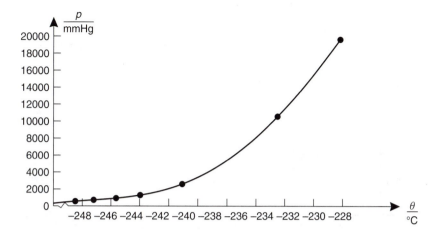

Figure 3.8   Graph of pressure against temperature for neon.

on the horizontal x-axis. This gives the graph shown in Figure 3.8, where we see that, as the temperature increases, the points become increasingly further apart, and so the plot has very uneven spaces between the points. This leads to significant differences in the extent of precision to which the position of each point can be defined. In particular, it makes it very difficult to make reasonable estimates of behaviour at small values of temperature and pressure. Suppose that instead of plotting pressure we take its logarithm to base 10 and plot that against temperature. Using a calculator gives us the following values:

$\dfrac{\theta}{°C}$	$\log\left(\dfrac{p}{mm\,Hg}\right)$
−228.7	4.297
−233.6	4.002
−240.2	3.501
−243.7	3.157
−245.7	2.912
−247.3	2.687
−248.5	2.512

Notice that the

$$\log\left(\frac{p}{mm\,Hg}\right)$$

values now occupy a much smaller range. From Figure 3.9, we can also see that they are far more evenly spread. This means that it is much easier to read off values at lower temperatures and pressures, and consequently we can predict the behaviour of neon more reliably.

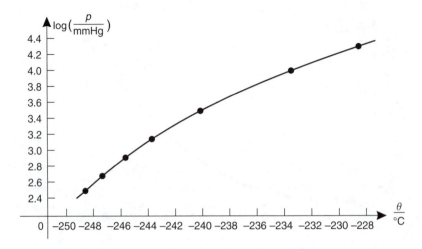

Figure 3.9   Graph of logarithm of pressure against temperature for neon.

3.6.2   The equation of a straight line

We saw in Section 3.4.2 on page 61 that a polynomial of order 1 was known as a linear expression. Some examples are

$$5x + 2$$

$$3x - 7$$

$$0.4x - 0.3$$

These can be summarized by the general equation

$$y = mx + c$$

where y is the value we evaluate on the right-hand side of the equals sign, and m and c are constants. So, if we write the first example as

$$y = 5x + 2$$

By convention the term intercept is used for the point where a line crosses the y-axis. There is also (in general) an intercept where the line crosses the x-axis, but this is of less use.

the constant m will be 5, and c will be 2. If we were to plot a graph of y against x, we would find that it was a straight line with gradient 5 and intercept 2, as shown in Figure 3.10. The consequence of this is that if we can rearrange any thermodynamic (or other mathematical) relationship into the form of a linear equation, we can obtain a straight line graph. Frequently, the values of the gradient and intercept of such graphs give us useful information about the system being studied.

Worked example 3.9

At the standard pressure $p^\circ = 1$ atm, a liquid boils at its normal boiling point, T_b. Given a set of values of the vapour pressure p at absolute temperature T, show how the Clausius–Clapeyron equation can be used to obtain the enthalpy of vaporization of the liquid.

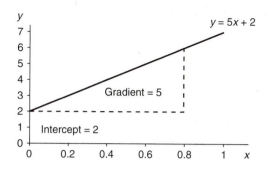

Figure 3.10 Graph of $y = 5x + 2$.

Chemical background

A liquid boils when its vapour pressure is equal to the atmospheric or external pressure. More molecules are then leaving the surface of the liquid than are reentering. Consequently, as the external pressure is reduced, the boiling temperature of the liquid falls.

Pressure is frequently expressed in a number of different units. Equivalent values for atmospheric pressure are 1 atm, 760 mm Hg, and 101.325 kPa.

Solution to worked example 3.9

Starting from the Clausius–Clapeyron equation

$$\ln\left(\frac{p_2}{p_1}\right) = \frac{\Delta_{vap}H}{R}\left[\left(\frac{1}{T_1}\right) - \left(\frac{1}{T_2}\right)\right]$$

we can substitute the pair of values given into the equation. It is simpler to set $p_1 = p^\circ$ and $T_1 = T_b$ and then to replace the symbols p_2 and T_2 by p and T, respectively. Doing this gives the expression

$$\ln\left(\frac{p}{p^\circ}\right) = \frac{\Delta_{vap}H}{R}\left[\left(\frac{1}{T_b}\right) - \left(\frac{1}{T}\right)\right]$$

At this stage, it is important to realize that p and T are the values of our variable pressure and temperature, respectively, whereas $\Delta_{vap}H$, R, p°, and T_b are all constants. If we expand the right-hand side of the equation (by removing the square brackets), we have

$$\ln\left(\frac{p}{p^\circ}\right) = \left(\frac{\Delta_{vap}H}{RT_b}\right) - \left(\frac{\Delta_{vap}H}{RT}\right)$$

which can be written as

$$\ln\left(\frac{p}{p^\circ}\right) = -\frac{\Delta_{vap}H}{R}\left(\frac{1}{T}\right) + \left(\frac{\Delta_{vap}H}{RT_b}\right)$$

Consider what happens if we make the substitutions

$$x = \frac{1}{T} \qquad\qquad y = \ln\left(\frac{p}{p^\circ}\right)$$

$$m = -\frac{\Delta_{\mathrm{vap}}H}{R} \qquad\qquad c = \frac{\Delta_{\mathrm{vap}}H}{RT_b}$$

You should find that this gives back the general equation

$$y = mx + c$$

If we plot $\ln\left(\dfrac{p}{p^\circ}\right)$ on the vertical y-axis and $1/T$ on the horizontal x-axis, a straight line will be obtained with gradient $-\dfrac{\Delta_{\mathrm{vap}}H}{R}$ and intercept $\dfrac{\Delta_{\mathrm{vap}}H}{RT_b}$.

Because the value of the gas constant R is known, the gradient gives the value of $\Delta_{\mathrm{vap}}H$. Notice that we need to plot $\ln\left(\dfrac{p}{p^\circ}\right)$ on the y-axis.

Because $p^\circ = 1$ atm, if the values of p are also expressed in atm, we will automatically be taking the natural logarithm of pure numbers, as required.

3.6.3   Plotting graphs

Having seen how to determine the plot we require, it is worth spending some time considering the mechanics of graph plotting. Although there are now many computer programs available for doing this, they can give unusual results and have difficulties with certain data sets. It is also worth discussing manual graph plotting because the rules that govern overall appearance should also be applied to computer generated plots.

If you have an equation relating two variables, you should be able to determine which variable to plot on the x- and y-axes. However, there will be times when you will be instructed to "plot variable 1 against variable 2." This is *always* equivalent to plotting "y against x," and it is often important to display them the right way around.

Having established this, the most important thing to remember is to fill as much of the graph paper as possible. It is best to have one plot per page, except when superimposing plots using the same axes for both, which can sometimes be useful. Try to avoid scales involving numbers such as 3 and 7 because these make it difficult to work out exactly where intermediate values should be plotted. Use a sharp pencil, label the axes and, if appropriate, include a title for the graph. There is no need to include the origin on either axis unless the spread of data lends itself to doing so.

Worked example 3.10

The vapour pressure p of water is given below as a function of temperature θ.

Determine the enthalpy of vaporization of water.

$\dfrac{\theta}{°C}$	$\dfrac{p}{10^{-2}\ \text{atm}}$
20	2.31
30	4.19
40	7.28
50	12.2
60	19.7
70	30.8

Chemical background

The vapour pressure increases with temperature as a result of inter-molecular forces in the liquid being broken. Vapour pressure measurements are consequently able to give us information about the nature of those forces. Most liquids with high values of enthalpy of vaporization also have high normal boiling points, although this is not always the case.

Typical values of the standard enthalpy of vaporization are 8.2 kJ mol^{-1} for methane and 43.5 kJ mol^{-1} for ethanol; the latter is higher due to the presence of hydrogen-bonding in the liquid.

Solution to worked example 3.10

We saw in the previous problem that we need to plot $\ln\left(\dfrac{p}{p^{o}}\right)$ against $1/T$

where $p^{o} = 1$ atm. It is important to realize that T refers to the absolute temperature (units of K) rather than the values given here. The first stage is to draw up a table containing the transformed data. If we consider the first pair of values, we have

$$\theta = 20°C$$

so that

$$T = (20 + 273) = 293\ \text{K}$$

and

$$\frac{1}{T} = \frac{1}{293\ K}$$

$$= 3.41 \times 10^{-3}\ \text{K}^{-1}$$

$$p = 2.31 \times 10^{-2}\ \text{atm}$$

so that

$$\frac{p}{p^o} = \frac{2.31 \times 10^{-2} \text{ atm}}{1 \text{ atm}}$$

$$= 2.31 \times 10^{-2}$$

and

$$\ln\left(\frac{p}{p^o}\right) = \ln(2.31 \times 10^{-2})$$

Powers of 10 are input to a

calculator using the button; negative powers also require use of

$$= -3.77$$

Applying a similar treatment to the other values leads to the final table:

$\dfrac{T^{-1}}{10^{-3}\,K^{-1}}$	$\ln\left(\dfrac{p}{p^o}\right)$
3.41	−3.77
3.30	−3.17
3.19	−2.62
3.10	−2.10
3.00	−1.63
2.92	−1.18

A suitable range of values on the x-axis (T^{-1}) would be from 2.9 to 3.5, with divisions of 0.1. On the y-axis ($\ln\left(\dfrac{p}{p^o}\right)$) it would be from −3.8 to −1.0, with divisions of 0.5. Exactly how these are fitted onto the graph paper depends on the particular arrangement of ruled squares. The labels on the axes are exactly as in the table above to ensure that the points on the graph actually represent pure numbers. To obtain the gradient, we need to measure the increase in y and divide it by the corresponding increase in x, making sure that we use as large a portion of the straight line as possible. Using the values shown on the graph in Figure 3.11, this gives us

$$\text{Gradient} = \frac{-1.18 - (-3.77)}{(2.92 - 3.41) \times 10^{-3}\,K^{-1}}$$

$$= \frac{2.59}{-4.9 \times 10^{-4}\,K^{-1}}$$

$$= -5.29 \times 10^3 \text{ K}$$

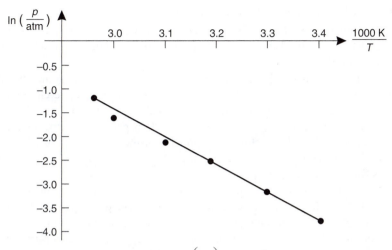

Figure 3.11 Graph of $\ln\left(\dfrac{p}{p^{\circ}}\right)$ against $\dfrac{1}{T}$ for water.

Notice that, in the first line of gradient calculation, the corresponding values of T^{-1} and $\ln\left(\dfrac{p}{p^{\circ}}\right)$ appear under one another. We know that the value of the gradient is equal to $-\dfrac{\Delta_{vap}H}{R}$, so

$$-\frac{\Delta_{vap}H}{R} = -5.29 \times 10^{3}\,K$$

Multiplying both sides of the equation by R and multiplying both sides by -1 to change the negative signs to positive signs on both sides gives

$$\Delta_{vap}H = 5.29 \times 10^{3}\ K \times R$$

which on substituting $R = 8.314$ J K^{-1} mol^{-1} gives

$$\Delta_{vap}H = 5.29 \times 10^{3}\ K \times 8.314\ \text{J K}^{-1}\ \text{mol}^{-1}$$

$$= 44.0 \times 10^{3}\ \text{J mol}^{-1}$$

This can be written more concisely by using kJ rather than 10^{3} J as

$$\Delta_{vap}H = 44.0\ \text{kJ mol}^{-1}$$

3.6.4   Plotting graphs using a spreadsheet

As in Chapter 2, Section 2.4.3.2, on page 45, the instructions given will refer to the Microsoft Excel spreadsheet; however, those for other spreadsheets will be very similar.

Begin by putting table headings at the top of the first two columns. In cell A1, type the letter q. Highlight the cell, select Format Cells, and choose the Symbol font. This changes the letter q to θ. In cell B1, type the letter p. We will leave a line for clarity, so the first value of θ, 20, is put into cell A3. Next to it in cell B3, type 2.31E-02; this is the spreadsheet shorthand for 2.31×10^{-2}. In a similar way we can enter the data to fill cells A4 to A8 with values of θ, and cells B4 to B8 with values of p.

The first step is to transform the data as we did manually in Section 3.6.3 on page 80. We will leave a blank column, so the transformed values of θ will be put into column D. Begin by putting the column heading T in cell D1. In D3 we type the formula =A3+273 to convert the temperature in °C to a value in K. Position the mouse at the bottom right corner of D3 and click and drag down to D8 to copy the formula. Now, we move to column E and put the heading 1/T in cell E1. The formula in cell E3 will be =1/D3; this needs to be copied by dragging the corner of E3 down as far as E8. In cell F1 we type the column heading "ln p"; in F3 the formula =LN(B3), which is copied by dragging to all the cells up to and including F8. The spreadsheet will now look as shown in Figure 3.12.

The data to be plotted now reside in columns E and F. Highlight cell E3 and drag the mouse down to E8 and across to F8, so that the cells containing data in both columns are highlighted. Choose the chart wizard icon (the coloured bar chart) from the tool bar, choose XY (scatter) and the icon for unconnected points. You can now click on Next, and choose appropriate labels for the axes and a title, to produce the final graph as shown in Figure 3.13.

Figure 3.12   Spreadsheet to illustrate data plotting.

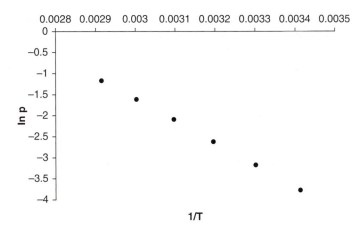

Figure 3.13 Spreadsheet showing plotted data.

3.7 The ideal gas equation

The ideal gas equation is the relationship you will probably meet most frequently in thermodynamics. This is because it describes the simplest form of matter to study, and is itself a simple expression. The volume of the molecules in an ideal gas would be negligible as would the forces between them. Real gases behave like ideal gases and obey the ideal gas equation at low pressures and high temperatures.

The ideal gas equation usually appears in the form

$$pV = nRT$$

where p is the pressure, V the volume, n the amount of gas, R the gas constant, and T the absolute temperature.

3.7.1 Interconversion of units

We have already met several examples in which units have been substituted along with the corresponding numerical value in an equation. This practice is essential and allows us to obtain the correct units of the final calculated quantity in a systematic way. A similar technique may also be used to convert a quantity from one set of units to another.

Worked example 3.11

The gas constant R is usually given as 8.314 J K^{-1} mol^{-1}. What is its value when expressed in the units dm^3 atm K^{-1} mol^{-1}?

Chemical background

The value of the gas constant was formerly determined from measurements of the molar volumes of oxygen and nitrogen, but more recently

We have already seen that for practical purposes we may wish to work with the pressure units of atm rather than kPa. Other non-SI units in common use are the Ångström (Å), which is equal to 10^{-10} m, and the calorie (cal), which is equal to 4.184 J. The use of such units in certain fields is so well established that to try to change would probably be more confusing than learning to live with them.

this has been done by determining the speed of sound in argon. Such changes in the way in which the values of constants are determined do take place from time to time, but invariably they lead to very small changes in their numerical values that are only significant in high-precision work.

Solution to worked example 3.11

To solve this problem, we need to use the following conversion factors:

$$1 \text{ atm} = 101.325 \text{ kPa} = 101.325 \times 10^3 \text{ Pa}$$

$$1 \text{ J} = 1 \text{ N m}$$

$$1 \text{ Pa} = 1 \text{ N m}^{-2}$$

Comparing the units of the two forms of R shows that we essentially have to convert J to dm^3 atm. We do this by successive substitution in the original expression for R.

$$R = 8.314 \text{ J K}^{-1} \text{ mol}^{-1}$$

$$= 8.314 \text{ N m K}^{-1} \text{ mol}^{-1}$$

Notice that, to obtain units of atm, we need to convert from Pa. At this stage, it helps to realize that

$$1 \text{ Pa} = 1 \text{ N m}^{-2} \text{ and } 1 \text{ J} = 1 \text{ N m}^{-2} \text{ m}^3$$

so that

$$R = 8.314 \text{ N m}^{-2} \text{ m } 3 \text{ K}^{-1} \text{ mol}^{-1}$$

$$= 8.314 \text{ Pa m}^3 \text{ K}^{-1} \text{ mol}^{-1}$$

Rearranging the definition for 1 atm and substituting 10 dm = 1 m gives us

$$R = \frac{8.314 \text{ atm}}{101.325 \times 10^3} \times (10 \text{ dm})^3 \text{ K}^{-1} \text{ mol}^{-1}$$

$$= 8.205 \times 10^{-5} \text{ atm} \times 10^3 \text{ dm}^3 \text{ K}^{-1} \text{ mol}^{-1}$$

$$= 8.205 \times 10^{-2} \text{ dm}^3 \text{ atm K}^{-1} \text{ mol}^{-1}$$

3.7.2 Constants and variables

The gas constant R is known as a *universal constant*, which means that its value does not change under any circumstances. This is in contrast to quantities such as n in the ideal gas equation. This is the amount of gas that will be fixed for a given sample of gas in a closed container but which may vary from sample to sample. The quantities p, V, and T are variables, but any of them may be considered to be fixed according to experimental conditions.

Worked example 3.12

Identify the constants and variables in the following equations that also describe the behaviour of gases.

(a) The van der Waals equation is

$$\left(p + \frac{an^2}{V^2} \right)(V - nb) = nRT$$

Here p is the pressure of gas, V its volume, T the temperature, and n the amount present. For carbon monoxide, $a = 0.1505$ Pa m^6 mol^{-2}, and $b = 0.0398 \times 10^{-3}$ m^3 mol^{-1}.

(b) The Beattie–Bridgeman equation is

$$p = \frac{RT\left[1 - \left(\dfrac{c}{V_m T^3} \right) \right]}{V_m^2}(V_m + B) - \frac{A}{V_m^2}$$

with

$$A = A_0 \left[1 - \left(\frac{a}{V_m} \right) \right]$$

and

$$B = B_0 \left[1 - \left(\frac{b}{V_m} \right) \right]$$

Here p is the pressure of gas, V_m its molar volume, and T the temperature. For methane, $A_0 = 0.230\ 71$ Pa m^6 mol^{-1}, $a = 18.55 \times 10^{-6}$ m^3 mol^{-1}, $B_0 = 55.87 \times 10^{-6}$ m^3 mol^{-1}, $b = -15.87 \times 10^{-6}$ m^3 mol^{-1}, $c = 12.83 \times 10$ m^3 mol^{-1}.

(c) The virial equation is

$$pV_m = RT + Bp$$

Here p is the pressure of gas, V_m its molar volume, R the gas constant, and T the absolute temperature. For nitrogen, the second virial coefficient B is -4.2 cm^3 mol^{-1} at 300 K.

Note that the constant B in (b) and (c) is different.

Chemical background

(a) The van der Waals equation was developed to account for the behaviour of a gas when it is liquefied. The volume term is reduced to allow for the finite volume of the molecules, whereas the pressure term is increased to allow for the presence of forces between molecules.

(b) The inclusion of a large number of parameters in the Beattie–Bridgeman equation allows its use in situations in which a precise fit to experimental data is required, particularly at high pressures.

(c) The form of the virial equation given here actually comes from a power series of the form

$$pV_m = RT + Bp + Cp^2 + Dp^3 + \ldots$$

where B, C, and D are functions of temperature and are known as the second, third, and fourth virial coefficients, respectively. One of the uses of this equation is that these coefficients may be related to the potential functions that describe the interaction between molecules. It is common to ignore terms in p^2 and higher.

Solution to worked example 3.12

(a) As in the case of the ideal gas equation, the variables are p, V, and T if the amount of gas n is fixed. The values of a and b are constant, depending only on the nature of the gas being considered.

(b) This is also similar to the ideal gas equation in that p, V_m, and T may vary. The other quantities A_0, a, B_0, b and c are constants and again depend only on the nature of the gas.

(c) As previously, p, V_m, and T are variables, and R is a constant. However, the question suggests that B is a function of temperature, so this could also be regarded as a variable, although its value at given temperatures may be obtained from tables.

3.7.3   *Proportion*

We can distinguish two types of proportion. If two quantities are *directly proportional* to one another, the ratio of their values will remain constant. For example, if x and y can have the following pairs of values, they are in direct proportion:

x	1	2	3	4
y	2	4	6	8

In this case the ratio y/x remains constant at 2. We say that y is directly proportional to x. This is written mathematically as

$$y \propto x$$

where $\propto$ is the proportionality sign. If this is the case, we can introduce a constant of proportionality K and write the relationship as

$$y = Kx$$

Rearranging this equation gives

$$K = \frac{y}{x}$$

and we see that the constant of proportionality is the constant ratio of the two quantities x and y.

Conversely, x and y may be *inversely proportional* to one another. An example of such a relationship is in the following table:

x	1	2	3	4
y	24	12	8	6

This time, the product xy remains constant at 24. In this case, we would write

$$y \propto \frac{1}{x}$$

which on replacing a by a new proportionality constant K' gives

$$y = \frac{K'}{x}$$

which rearranges to

$$K' = xy$$

so K' is the constant product of x and y.

Worked example 3.13

For a fixed amount of gas that obeys the ideal gas equation, describe the relationship between the two variables indicated when the third is held constant:

(a) p and V at constant T
(b) p and T at constant V
(c) T and V at constant p

Chemical background

The experimental relationships between these variables are expressed by Boyle's law and Charles's law. Together with Avogadro's law, these can be used to formulate the ideal gas equation. This problem involves deducing the original relationships from the final equation.

Solution to worked example 3.13

To solve this problem we need to be able to rearrange the ideal gas equation

$$pV = nRT$$

so that the two variables are on opposite sides of the equation. In each case, we will call the grouping of constant terms K, which will be our constant of proportionality.

(a) Dividing both sides of the equation by V gives

$$p = \frac{nRT}{V}$$

and because nRT is a constant, which we will call K, we can write

$$p = \frac{K}{V}$$

or

$$p \propto \frac{1}{V}$$

so p and V are inversely proportional to one another. This is Boyle's law.

(b) Using the rearranged equation from part (a)

$$p = \frac{nRT}{V}$$

and because nR/V is a constant, which we will call K', we can write

$$p = K'T$$

or

$$p \propto T$$

so p and T are directly proportional.

(c) Dividing both sides of the original equation by p gives

$$V = \frac{nRT}{p}$$

and because nR/p is a constant, which we will call K'', we can write

$$V = K''T$$

or

$$V \propto T$$

so V and T are directly proportional. This is Charles's law.

3.7.4   Functions of two variables

In Section 3.4.3 on page 64 we met the idea of writing an expression as a function of a variable, using the $f(x)$ notation. This can be extended to cases where we have two (or more) variables, such as

$$f(x, y) = x^3 + 3x^2y + 2y^2$$

and

$$g(x, y) = \ln(3x) + 2xy$$

Note that we can also write expressions such as $z = f(x,y)$ and $z = g(x,y)$ in order to facilitate the plotting of such functions in three dimensions.

Worked example 3.14

Use the function notation to write expressions for the

(a) pressure
(b) volume
(c) temperature

of an ideal gas.

Chemical background

Although we can display a function of one variable as a line graph, when two variables are involved, this is not possible. One solution is to use a three-dimensional plot, where the height above the axes gives the value of the functions. This has been done for this example in Figure 3.14.

Solution to worked example 3.14

(a) We can use the rearranged equation from part (a) of worked example 3.12. Because both temperature and volume are variables, we write

$$p(T,V) = \frac{nRT}{V}$$

(b) We can use the rearranged equation from part (c) of worked example 3.12. Because both temperature and pressure are variables, we write

$$V(T,p) = \frac{nRT}{p}$$

(c) We can rearrange the ideal gas equation (by dividing both sides by nR) to make T the subject:

$$T = \frac{pV}{nR}$$

Because p and V are both variables,

$$T(p,V) = \frac{pV}{nR}$$

Figure 3.14 Plots representing the ideal gas equation. Units are Pa for pressure p, m^3 for volume V, and K for absolute temperature T.

3.7.5   Partial differentiation

We have already met the idea of differentiating a function of one variable in Section 3.5.1 on page 66. We saw that this was equivalent to determining the gradient of the curve representing that function, and represented the change in the function for a given change in the variable.

When we have a function of two variables, changing either of them will produce a change in the value of the function. This change is given by a quantity called the *partial derivative*. If our function is $f(x, y)$, then it is possible to determine

$$\left(\frac{\partial f(x, y)}{\partial x} \right)_y$$

the partial derivative with respect to x when y is held constant, and

$$\left(\frac{\partial f(x, y)}{\partial y} \right)_x$$

the partial derivative with respect to y when x is held constant. Note that, in both cases, the quantity to be held constant is shown outside the bracket. The notation for each derivative is often abbreviated to

$$\frac{\partial f}{\partial x} \quad \text{and} \quad \frac{\partial f}{\partial y}$$

respectively, for clarity and to save space.

The quantity $\dfrac{\partial f}{\partial x}$ represents the change in $f(x, y)$ when x is changed, and is equal to the gradient of the tangent to the surface of $f(x, y)$ in the x-direction, as shown in Figure 3.15 in which the tangent is drawn at a constant value of y, denoted as k. Similarly, $\dfrac{\partial f}{\partial y}$ represents the change in $f(x, y)$ when y is changed, and is equal to the gradient of the tangent to the surface $f(x, y)$ in the y-direction.

Partial derivatives are calculated in the same way as full derivatives, while holding the fixed variable constant. If a function is defined so that

$$f(x, y) = 2x^2 + 3xy + 2xy^2$$

we can rewrite this treating every term in y as a constant. In the expression below, all the constant terms are enclosed in brackets:

$$f(x, y) = (2)x^2 + (3y)x + (2y^2)x$$

Differentiating x^2 gives $2x$, and differentiating x gives 1, so we obtain

$$\frac{\partial f}{\partial x} = 2(2x) + (3y) \times 1 + (2y^2) \times 1$$

$$= 4x + 3y + 2y^2$$

Similarly, if we rewrite the expression for $f(x,y)$ treating every term in x as a constant, then

$$f(x, y) = (2x^2) + (3x)y + (2x)y^2$$

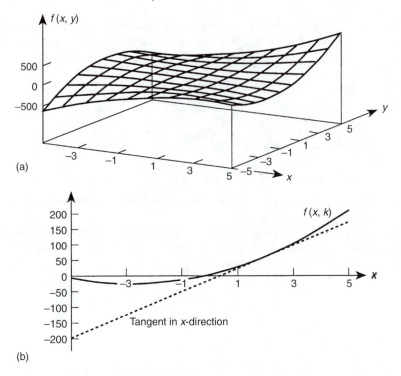

Figure 3.15   The partial derivative $\dfrac{\partial f(x, y)}{\partial x}$ at $y = k$.

Differentiating a constant gives 0, differentiating y gives 1, and differentiating y^2 gives $2y$. We therefore obtain

$$\frac{\partial f}{\partial y} = 0 + (3x) \times 1 + (2x)2y$$

$$= 3x + 4xy$$

Worked example 3.15

For an ideal gas, calculate the partial derivatives

$$\frac{\partial p}{\partial V} \quad \text{and} \quad \frac{\partial p}{\partial T}$$

Chemical background

Partial derivatives find much use in thermodynamics, as it is possible to derive relationships between them. One example of this is Maxwell's relations, which consist of four relationships of the type

$$\left(\frac{\partial V}{\partial T}\right)_p = -\left(\frac{\partial S}{\partial p}\right)_T$$

These are useful because, although it is easy to measure the variation of volume V with temperature T, it is not straightforward to measure entropy S as a function of pressure p in the laboratory.

The other Maxwell relations are:

$$\left(\frac{\partial T}{\partial V}\right)_S = -\left(\frac{\partial p}{\partial S}\right)_V$$

Solution to worked example 3.15

We can use the expression obtained in part (a) of worked example 3.14:

$$\left(\frac{\partial T}{\partial p}\right)_S = \left(\frac{\partial V}{\partial S}\right)_p$$

To calculate the partial derivative with respect to V, we treat T as a constant, so the expression on the right-hand side can be written with the constant term in brackets as

$$\left(\frac{\partial p}{\partial T}\right)_V = \left(\frac{\partial S}{\partial V}\right)_T$$

$$p(T,V) = (nRT)\frac{1}{V}$$

$$= (nRT)V^{-1}$$

Remembering that we differentiate by multiplying by the power and reducing the power by 1, x^{-1} would differentiate (with respect to x) to give $(-1)x^{-2}$, i.e.,

$$\frac{d}{dx}(x^{-1}) = (-1)x^{-1-1} = -\frac{1}{x^2}$$

Therefore we obtain

$$\frac{\partial p}{\partial V} = nRT\left(-\frac{1}{V^2}\right)$$

$$= -\frac{nRT}{V^2}$$

To calculate the partial derivative with respect to T, we treat V as a constant, so grouping the constant term in brackets gives us

$$p(T,V) = \left(\frac{nR}{V}\right)T$$

Remember that differentiating x with respect to x gives us 1, i.e.,

$$\frac{dx}{dx} = 1$$

Here we are differentiating with respect to T, which gives

$$\frac{\partial p}{\partial T} = \left(\frac{nR}{V} \right) \times 1$$

$$= \frac{nR}{V}$$

3.7.6 The differential

We saw, in worked example 3.10, that we could determine the gradient m of a straight line graph of y against x as

$$m = \frac{\text{increase in } y}{\text{increase in } x} = \frac{\Delta y}{\Delta x}$$

It is possible to rearrange this equation to give an expression for the increase in either of the variables x or y:

$$\Delta x = \frac{\Delta y}{m}, \Delta y = m\Delta x$$

Similarly, we can deduce from Section 3.5.1 on page 66 that the gradient of a curve plotting y against x is given by the derivative dy/dx, which comes from considering very small or infinitesimal changes in the variables x and y. By analogy with our treatment for the straight line above, we can write

$$dy = m \, dx$$

where the gradient m will be a function of x. This expression allows us to express the change in variable y in terms of the variable x and the change in x.
For example, if

$$y = 2x^2 + x + 3$$

the derivative

$$\frac{dy}{dx} = 4x + 1$$

and so

$$dy = (4x + 1) \, dx$$

where dy is known as the *differential*.
This approach can be readily extended to functions of more than one variable, such as $f(x, y)$. The differential $df(x, y)$ is then defined by

$$df(x, y) = \left(\frac{\partial f}{\partial x} \right)_y dx + \left(\frac{\partial f}{\partial y} \right)_x dy$$

This can be interpreted as giving the change in the function $f(x, y)$ when the variables x and y are changed by small amounts dx and dy, respectively. For example, if

$$f(x, y) = xy^3 + 2x^2y$$

we calculate the partial derivative $\dfrac{\partial f}{\partial x}$ by treating y as a constant. Then

$$f(x,y) = (y^3)x + (2y)x^2$$

and

$$\frac{\partial f}{\partial x} = (y^3) \times 1 + (2y)2x$$

$$= y^3 + 4xy$$

Treating x as a constant allows us to calculate the partial derivative $\dfrac{\partial f}{\partial y}$:

$$f(x, y) = (x)y^3 + (2\,x^2)y$$

and

$$\frac{\partial f}{\partial y} = (x)3y^2 + (2x^2) \times 1$$

$$= 3xy^2 + 2x^2$$

Substituting in our general expression given earlier for the differential gives:

$$df = (y^3 + 4xy)\,dx + (3xy^2 + 2x^2)\,dy$$

Note that for simplicity we have replaced the term $df(x,y)$ by df.

Worked example 3.16

Calculate the differential dp of an ideal gas.

Chemical background

Differentials can be exact or inexact. For an exact differential, the relationship

$$\left[\frac{\partial}{\partial y} \left(\frac{\partial f}{\partial x} \right)_y \right]_x = \left[\frac{\partial}{\partial x} \left(\frac{\partial f}{\partial y} \right)_x \right]_y$$

must hold. If this is the case, the function $f(x,y)$ is known as a state function. This means that its value is independent of the path taken to get to the particular state. This idea was discussed more fully in Section 3.3 on page 56.

Solution to worked example 3.16

We first need to write the pressure p of the ideal gas as a function of the other variables by rearranging the ideal gas equation. Because $pV = nRT$, we have

$$p = \frac{nRT}{V}$$

Because the variables on the right-hand side of this equation are the temperature T and the volume V, it may be clearer to write this using our notation for a function of two variables:

$$p(T,V) = \frac{nRT}{V}$$

To determine the differential dp, we need to calculate the partial derivatives

$$\left(\frac{\partial p}{\partial T}\right)_V \quad \text{and} \quad \left(\frac{\partial p}{\partial V}\right)_T$$

These were determined in worked example 3.15, and were

$$\left(\frac{\partial p}{\partial T}\right)_V = \frac{nR}{V} \quad \text{and} \quad \left(\frac{\partial p}{\partial V}\right)_T = -\frac{nRT}{V^2}$$

The differential dp will be given by the expression

$$dp = \left(\frac{\partial p}{\partial T}\right)_V dT + \left(\frac{\partial p}{\partial V}\right)_T dV$$

so we can substitute directly and obtain

$$dp = \left(\frac{nR}{V}\right)dT - \left(\frac{nRT}{V^2}\right)dV$$

It is relatively straightforward to show that

$$\left[\frac{\partial}{\partial V}\left(\frac{\partial p}{\partial T}\right)_V\right]_T = \left[\frac{\partial}{\partial T}\left(\frac{\partial p}{\partial V}\right)_T\right]_V = -\frac{nR}{V^2}$$

confirming that dp is an exact differential.

3.8   The van der Waals equation

We have already met the van der Waals equation to describe the behaviour of a gas in worked example 3.12, in which we saw that corrections were made to both the pressure and volume to improve the description of real gases. These are the constants a and b in the equation

$$\left(p + \frac{an^2}{V^2}\right)(V - nb) = nRT$$

in which all the other symbols have the same meaning as in the ideal gas equation.

3.8.1   Expansion of brackets

The mathematics involved in the expansion of expressions using brackets has been discussed in Section 3.4.1 on page 59. We saw there that

$$(a + b)(c + d) = ac + ad + bc + bd$$

Worked example 3.17

Remove the brackets from the van der Waals equation.

Chemical background

The constants a and b in the van der Waals equation are normally obtained by fitting the equation to experimental measurements of pressure, volume, and temperature.

Alternatively, because there is a critical temperature T_c above which a gas cannot be liquefied by pressure alone, it is possible to define a corresponding critical pressure p_c and critical volume V_c. The quantities are related to the van der Waals constants by the equations

$$a = 3p_c V_c^2 \quad \text{and} \quad b = \frac{V_c}{3}$$

Solution to worked example 3.17

Multiplying the second bracket by each of the terms in the first bracket gives

$$p(V - nb) + \left(\frac{an^2}{V^2}\right)(V - nb) = nRT$$

Each term inside the remaining brackets of $(V - nb)$ can now be multiplied by the term outside to give

$$pV - pnb + \left(\frac{an^2}{V^2}\right)V - \left(\frac{an^2}{V^2}\right)nb = nRT$$

We can now simplify the third and fourth terms on the left-hand side of this equation:

$$\left(\frac{an^2}{V^2}\right)V = \frac{an^2V}{V^2} = \frac{an^2}{V} \quad \text{and} \quad \left(\frac{an^2}{V^2}\right)nb = \frac{an^3b}{V^2}$$

because

$$\frac{V}{V^2} = V^{1-2} = V^{-1} = \frac{1}{V} \quad \text{and} \quad n^2 \times n = n^{2+1} = n^3$$

so that the overall equation becomes

$$pV - pnb + \frac{an^2}{V} - \frac{an^3b}{V^2} = nRT$$

This would be a reasonable way of leaving the equation, as the brackets have been removed and all the terms simplified. However, a more elegant solution would be to continue as follows. The nRT term on the right can be "removed" by subtracting nRT from both sides of the equation to give

$$pV - pnb + \frac{an^2}{V} - \frac{an^3b}{V^2} - nRT = 0$$

If we now multiply each term by V^2 (noting that $0 \times V^2 = 0$), we obtain

$$pV^3 - pnbV^2 + an^2V - an^3b - nRTV^2 = 0$$

or

$$pV^3 - (pnb + nRT)V^2 + an^2V - an^3b = 0$$

which can be classified as a cubic expression in V and contains no fractions.

3.8.2   Combining limits

The concept of a limit was discussed in Section 3.5.1 on page 66. Limits can be combined in much the same way as the quantities they represent. For example, if

$$f(x) = g(x) + h(x)$$

then

$$\operatorname*{Lim}_{x \to L} f(x) = \operatorname*{Lim}_{x \to L} g(x) = \operatorname*{Lim}_{x \to L} h(x)$$

Similarly, if

$$f(x) = g(x)\, h(x)$$

then

$$\operatorname*{Lim}_{x \to L} f(x) = \operatorname*{Lim}_{x \to L} g(x) = \operatorname*{Lim}_{x \to L} h(x)$$

Worked example 3.18

Show that the van der Waals equation reduces to the ideal gas equation in the limit of low pressure.

Chemical background

A general requirement of all acceptable equations of state for gases is that they will reduce to the ideal gas equation as the pressure falls to zero. The interactions between molecules are then negligible, as is the volume of the molecules relative to the volume of the gas, so we are approaching the ideal situation.

Solution to worked example 3.18

In terms of limits, we can write the van der Waals equation as

$$\operatorname*{Lim}_{p \to 0}\left(p + \frac{an^2}{V^2} \right)\operatorname*{Lim}_{p \to 0}(V - nb) = \operatorname*{Lim}_{p \to 0} nRT$$

The first term to consider is

$$p + \frac{an^2}{V^2}$$

in the first bracket. We need to realize that, as the pressure p tends towards zero, the volume V tends towards infinity. This can be expressed mathematically as $p \to 0$, $V \to \infty$.

As V becomes very large, $\dfrac{an^2}{V^2}$ will tend towards zero. Also because the V term is squared, this will happen faster than p tends to zero. Thus this term will tend to $p+0$, or just p. Mathematically, we can express the steps in this argument as

$$\lim_{p \to 0} \left(p + \frac{an^2}{V^2} \right) = \lim_{p \to \infty} \left(p + \frac{an^2}{V^2} \right) = p$$

Now we consider the term in the second bracket. Again, as $p \to 0$, $V \to \infty$ so

$$\lim_{p \to 0} (V - nb) = \lim_{p \to \infty} (V - nb)$$

As V becomes very large, the quantity nb remains the same, and so the value of the second limit becomes closer to V:

$$\lim_{p \to 0} (V - nb) = V$$

Finally, as neither p nor V appears on the right-hand side of the equation, we have

$$\lim_{p \to 0} (nRT) = nRT$$

Substituting into our equation involving limits, this now gives us

$$pV = nRT$$

which is the ideal gas equation, as required.

Some of the arguments used above for determining each individual limit may seem difficult to replicate in different situations. Limits generally arise in the derivation of equations, so it is usually sufficient to be able to follow these arguments rather than developing them from scratch.

3.9   Equilibrium constants

The equilibrium constant is a fundamental quantity in thermodynamics, describing the equilibrium state of a system in terms of concentration units. Its magnitude determines the extent to which a particular reaction will take place; a large value indicates a greater concentration of products than reactants and hence a favourable forward reaction.

A mathematical way of defining the equilibrium constant K is to use the equation

$$K = \prod_j a_j^{v_j}$$

The symbol a_j here is used to represent the activity of species j; activities are the correct way of defining dimensionless equilibrium constants, i.e., those without units. The activity a_j of species j having concentration c_j can thus be defined by

$$a_j = \left(\frac{c_j}{c^o}\right)\gamma_j$$

where c^o is the standard concentration taken as 1 mol dm^{-3} exactly, and γ_j is the activity coefficient of species j.

The mathematical symbol $\prod$ is new to us. It simply means "multiply," or "take the product." The symbol v_j is known as the stoichiometric number of species j. This is obtained by rewriting our general chemical reaction

$$a\text{A} + b\text{B} + \ldots \rightarrow c\text{C} + d\text{D} + \ldots$$

as

$$c\text{C} + d\text{D} + a\text{A} - b\text{B} - 0$$

where the reactants have been subtracted from the products. We then have

$$\text{Stoichiometric number of A} = v_\text{A} = -a$$

$$\text{Stoichiometric number of B} = v_\text{B} = -b$$

$$\text{Stoichiometric number of C} = v_\text{C} = c$$

$$\text{Stoichiometric number of D} = v_\text{D} = d$$

Worked example 3.19

Calculate the equilibrium constant for the dissociation of ethanoic acid at 25°C if the equilibrium concentrations are

$$[\text{CH}_3\text{COOH}] = 0.90 \text{ mol dm}^{-3}$$

and

$$[\text{H}^+] = [\text{CH}_3\text{COO}^-] = 0.004 \text{ mol dm}^{-3}$$

Chemical background

Ethanoic acid is a weak electrolyte, and we can see from the concentration values just given that the concentration of undissociated acid is far higher than that of the dissociated ions. We therefore expect to obtain an equilibrium constant value that is considerably less than 1. When the equilibrium constant is applied to acid dissociations such as this, it is usually given the symbol K_a. This leads to the use of logarithms for expressing acid strength in terms of pK_a, which is defined as

$$pK_a = -\log K_a$$

Ethanoic acid is more commonly known industrially by its older name of acetic acid. Along with acetic anhydride, it is produced by the catalytic oxidation of acetaldehyde and is used in the production of cellulose acetate, which is needed for the manufacture of upholstery and carpets.

where log represents the logarithm taken to base 10, as we saw in Section 3.6.1 on page 75.

Solution to worked example 3.19

The first step in this problem is to write a balanced chemical equation for the dissociation of ethanoic acid. The species required are all given in the question, and the required equation is

$$CH_3COOH = CH_3COO^- + H^+$$

Subtracting the reactant from the ethanoic acid product gives

$$CH_3COO^- + H^+ - CH_3COOH = 0$$

and the stoichiometric numbers are therefore

$$\nu_{CH_3COO^-} = 1$$

$$\nu_{H^+} = 1$$

$$\nu_{CH_3COOH} = -1$$

Substituting into the general expression for the equilibrium constant then gives

$$K = \left(\frac{c_{CH_3COO^-}}{c^o} \right)^1 \left(\frac{c_{H^+}}{c^o} \right)^1 \left(\frac{c_{CH_3COOH}}{c^o} \right)^{-1}$$

where c denotes the concentration of each species. As in this case, we often assume that the activity coefficients are all close to unity.

Because

$$\left(\frac{c_{CH_3COOH}}{c^o} \right)^{-1} = \frac{1}{c_{CH_3COOH} \Big/ c^o}$$

this can be rewritten as

$$K = \frac{\left(\dfrac{c_{CH_3COO^-}}{c^o} \right) \left(\dfrac{c_{H^+}}{c^o} \right)}{\left(\dfrac{c_{CH_3COOH}}{c^o} \right)}$$

The values given in the question can now be substituted into this expression:

$$K = \frac{0.004 \times 0.004}{0.90} = 1.8 \times 10^{-5}$$

Notice that the use of c° means that the units of mol dm^{-3} do not need to appear in the equation for calculating K.

3.9.1   Solving quadratic equations using a formula

In Section 3.4.2 on page 61, we saw that a polynomial expression of order 2 is called a quadratic. A quadratic equation is one of the form

$$ax^2 + bx + c = 0$$

where a, b, and c are constants. Examples are

$$3x^2 + 2x - 1 = 0$$

and

$$x^2 + 6x + 4 = 0$$

The graphs of these equations are shown in Figure 3.16, in which the solutions are given by the intersection of the curves with the x-axis, i.e., when the value of the expression is zero.

This may happen at none, one, or two values of x. In some cases, these can be found by factorizing the expression into two brackets and setting each in turn to zero. However, this is not always possible. Another method that is more generally applicable is to apply the standard formula.

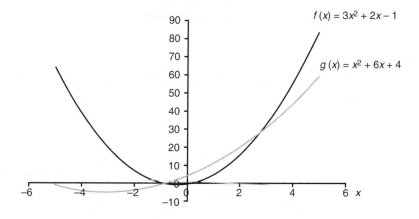

Figure 3.16   Graphs of the quadratic functions $f(x) = x^2 + 2x - 1$ and $g(x) = x^2 + 6x + 4$.

The solution of the general quadratic equation given at the start of this section is

$$x = \frac{-b \pm \sqrt{b^2 - 4ac}}{2a}$$

Notice that the $\pm$ sign ensures that two values of x will result; however, solutions will only be found if b^2 is greater than the quantity $4ac$, because otherwise we would be taking the square root of a negative number. We will return to this topic again in Chapter 6.

Worked example 3.20

In the reaction between nitrogen and hydrogen to give ammonia, the fractional amount of nitrogen reacted x can be related to the equilibrium constant K, which has the value 977. This gives the quadratic equation

$$81.2x^2 - 163.4x + 81.2 = 0$$

for x. What is the value of x?

Chemical background

Ammonia is manufactured industrially by means of the Haber process. This involves the reaction

$$N_{2(g)} + 3H_{2(g)} = 2NH_{3(g)}$$

which takes place at 450°C and 200 atmospheres in the presence of an iron catalyst. These conditions are used to give an acceptable reaction rate; the equilibrium constant gives us no information about this. Nitrogen is obtained from the air, whereas hydrogen is usually obtained from the reaction of natural gas with steam. The Haber process is actually a major use for hydrogen gas.

Ammonia is also obtained commercially as a by-product of the generation of coke from coal. It is contained in an aqueous phase and is liberated by means of steam distillation. The largest use of ammonia is in fertilizers, but it is also used on a large scale in the production of nitric acid.

Solution to worked example 3.20

Relating the above expression to the general equation

$$ax^2 + bx + c = 0$$

gives $a = 81.2$, $b = -163.4$, and $c = 81.2$. Substituting these values into the equation for x

$$x = \frac{-b \pm \sqrt{b^2 - 4ac}}{2a}$$

gives

$$x = \frac{-(-163.4) \pm \sqrt{(-163.4)^2 - 4 \times 81.2 \times 81.2}}{2 \times 81.2}$$

This can be solved in stages using a calculator to give

$$x = \frac{163.4 \pm \sqrt{26699.6 - 26373.8}}{162.4}$$

$$= \frac{163.4 \pm \sqrt{325.8}}{162.4}$$

$$= \frac{163.4 \pm 18.05}{162.4}$$

so

$$x = 1.12 \quad \text{or} \quad x = 0.895$$

We have, as expected, obtained two values for x. However, if we look back at the question, we see that only one is required. How do we know which one to take? If we look closely, we see that x is defined as the fractional amount of nitrogen reacted. Its value must therefore be less than 1, so we take $x = 0.895$ as our solution.

We use the $\boxed{x^2}$ button to square the value of b, and then the $\boxed{\sqrt{\ }}$ button to take the required square root.

It is often necessary to eliminate one of the mathematical solutions in this way. Sometimes a negative and a positive value will be obtained, with the negative value obviously not making sense.

3.9.2 Solving quadratic equations iteratively

Although the equation introduced in the previous section is tedious to apply, it will give the solution to a quadratic equation without the need to understand any difficult mathematical concepts. However, it is possible to obtain the desired solution without using this formula.

Begin by remembering that x is a positive fractional amount, so its value must lie between 0 and 1. Because

$$f(x) = 81.2x^2 - 163.4x + 81.2$$

we have

$$f(0) = 81.2 \times 0^2 - 163.4 \times 0 + 81.2 = 81.2$$

$$f(0.5) = 81.2 \times 0.5^2 - 163.4 \times 0.5 + 81.2 = 19.8$$

$$f(1.0) = 81.2 \times 1.0^2 - 163.4 \times 1.0 + 81.2 = -1.00$$

so that our solution must lie between $x = 0.5$ and $x = 1.0$ because $f(x) = 0$ represents the point where the curve crosses the x-axis.

Use the $\boxed{x^2}$ button to work out each square.

In a similar fashion we determine

$$f(0.75) = 81.2 \times 0.75^2 - 163.4 \times 0.75 + 81.2 = 4.33.$$

This is positive, so our solution lies between $x = 0.75$ and $x = 1.0$. The halfway point between these is at $x = 0.875$, so we calculate

$$f(0.875) = 81.2 \times 0.875^2 - 163.4 \times 0.875 + 81.2 = 0.39$$

This is still positive, so our solution lies between $x = 0.875$ and $x = 1.0$. The midpoint is at $x = 0.9375$, so

$$f(0.9375) = 81.2 \times 0.9375^2 - 163.4 \times 0.9375 + 81.2 = -0.62$$

Because this is negative, our solution lies between $x = 0.875$ and $x = 0.9375$. The midpoint of these is $x = 0.906$

$$f(0.906) = 81.2 \times 0.906^2 - 163.4 \times 0.906 + 81.2 = -0.19$$

giving a root between $x = 0.875$ and $x = 0.906$. We once again determine the midpoint of these, which is $x = 0.891$.

$$f(0.891) = 81.2 \times 0.891^2 - 163.4 \times 0.891 + 81.2 = 0.07$$

So the root is between $x = 0.891$ and $x = 0.906$. We determine $f(x)$ at the midpoint, $x = 0.899$

$$f(0.899) = 81.2 \times 0.899^2 - 163.4 \times 0.899 + 81.2 = -0.07$$

indicating a root between $x = 0.891$ and $x = 0.899$. The function at the midpoint $x = 0.895$ is

$$f(x) = 81.2 \times 0.895^2 - 163.4 \times 0.895 + 81.2 = 0.00$$

The root is therefore taken as $x = 0.895$, as expected from our previous solution of this problem.

Note that this procedure only needs to be continued until the two values of x agree to within the desired degree of precision.

Exercises

1. Rearrange the following equations to make y the subject:

(a) $x + y = 7$

(b) $9x - 3y = 18$

(c) $xy = 12$

2. Rearrange the following equations to make y the subject:

(a) $\dfrac{1}{x} - \dfrac{1}{y} = 1$

(b) $4xy^2 = 16$

(c) $10 - 2xy = 0$

3. Remove the brackets from the following expressions:

   (a) $xy(x^2 + y^2)$

   (b) $x(x + y)(x - y^2)$

   (c) $(x + y)(x - y)$

4. For each of the following functions:

   (i)   Write down the order of the expression.

   (ii) Calculate $f(-3)$.

   (iii) Write down the derivative $\dfrac{df(x)}{dx}$.

      (a) $f(x) = 3x^4 + 2x^3 + 4x^2 + x + 6$

      (b) $f(x) = 8x^6 + x^5$

      (c) $f(x) = 8x^3 + 1$

5. Rewrite the expression for

$$f(x) = \frac{1}{x} - \frac{2}{x^2} + \frac{3}{x^3} - \frac{4}{x^4} + \frac{5}{x^5}$$

   in terms of powers of x, and hence calculate $\dfrac{df(x)}{dx}$.

6. Give the values of the following quantities:

   (a)   log 24

   (b)   log 10

   (c)   ln 3

7. Give the gradients and intercepts of the straight lines represented by the following equations:

   (a) $y = 6x + 3$

   (b) $3y = 5x + 4$

   (c) $x - 2y = 8$

8. For the following data, plot ln y against $1/x$ using a pencil and graph paper:

x	1.0	2.0	3.0	4.0	5.0
y	0.368	0.135	0.097	0.082	0.074

9. For the following data, plot $1/y$ against x^2 using a spreadsheet graphing function:

x	−3.0	−2.0	−1.0	0.0	1.0	2.0	3.0
y	0.0377	0.0869	0.400	−2.00	0.400	0.0869	0.0377

10. If x and y are directly proportional, and x takes the value 10 when y is 30, what is the value of the proportionality constant?

11. If x and y are inversely proportional, and x takes the value 6 when y is 3, what is the value of x when y is 10?

12. State the relationship between x and y if they can take the following pairs of values:

x	64	32	16	8	4	2	1
y	1	2	4	8	16	32	64

13. If $f(x, y) = 3x^2y - 2xy$, calculate

(a) $f(1,-1)$

(b) $f(0, 2)$

(c) $f(-2, 1)$

14. If $f(x, y) = \dfrac{1}{x} - \dfrac{1}{y}$, calculate

(a) $f(-1,-1)$

(b) $f(1,1)$

(c) $f(1,-2)$

When is $f(x, y)$ not defined?

15. If $f(x, y) = xy + \ln(xy)$, calculate

(a) $f(1,1)$

(b) $f(1,2)$

(c) $f(1,3)$

For what values of x and y is $f(x, y)$ not defined?

16. If $f(x, y) = 3x^2 + 2xy + 4x^3y^2$ calculate

$$\frac{\partial f}{\partial x} \quad \text{and} \quad \frac{\partial f}{\partial y}$$

and write an expression for the differential $df(x, y)$.

17. Show that the $df(x, y)$ in Question 16 is an exact differential.

18. If $f(x, y) = x^2 + x^2y + xy^2 + y^2$, calculate

$$\frac{\partial f}{\partial x} \quad \text{and} \quad \frac{\partial f}{\partial y}$$

19. Solve these quadratic equations using the formula:

(a) $x^2 + 9x - 7 = 0$

(b) $4x^2 + 7x + 2 = 0$

(c) $0.20x^2 - 0.90x + 0.20 = 0$

20. The quadratic equation

$$0.24x^2 + 0.20x - 0.16 = 0$$

has one root between $x = -1.5$ and $x = -1.0$, and another between $x = 0.45$ and $x = 0.55$. Find these roots using an iterative trial and error method.

Problems

1. Rearrange each of the following equations to make the given variable the subject:

(a) $\Delta G = \Delta H - T\Delta S$ (subject T)

(b) $V = n_1V_1 + n_2V_2$ (subject V_1)

(c) $E = \dfrac{n^2h^2}{8ma^2}$ (subject n)

2. If a gas obeys the ideal gas equation

$$pV = nRT$$

state the value of the gradient obtained when each of the following plots were made:

(a) p against T at constant n and V

(b) p against n at constant V and T

(c) V against n at constant p and T.

3. A simplified representation of the heat capacity C_p in terms of the absolute temperature T is given by

$$C_p = a + bT$$

where a and b are constants. If C_p is measured at various values of T, how can the values of a and b be determined?

4. The standard change $\Delta G°$ in the Gibbs free energy for an equilibrium reaction is related to its equilibrium constant K by the equation

$$\Delta G° = -RT \ln K$$

Use the graphing function on a spreadsheet to plot $\Delta G°$ against K for values of K ranging from 10^{-4} to 10^4 at 298 K with $R = 8.31$ J K^{-1} mol^{-1}.

5. The equilibrium constant K is related to the standard enthalpy of reaction $\Delta H°$ and the standard entropy of reaction $\Delta S°$ by the equation

$$\ln K = -\left(\frac{\Delta H°}{RT}\right) + \left(\frac{\Delta S°}{R}\right)$$

where R is the gas constant and T is the absolute temperature. Calculate

$$\frac{d(\ln K)}{dT}$$

assuming that $\Delta H°$ and $\Delta S°$ are independent of temperature.

6. We saw in the chapter that the heat capacity C_p at constant pressure can be represented by the equation

$$C_p = a + bT + \frac{c}{T^2}$$

Obtain an expression for $\dfrac{dC_p}{dT}$ in terms of a, b, c, and T.

7. Derive an expression for the differential dV for a gas that obeys the ideal gas equation

$$pV = nRT$$

assuming that n is a constant.

8. The van der Waals constants for oxygen used to be given as $a = 1.36$ atm l^2 mol^{-2} and $b = 0.0318$ l mol^{-1}. Use the conversion factors below to express these quantities in the appropriate SI units, which are Pa m^6 mol^{-2} and m^3 mol^{-1}, respectively.

$$1 \text{ atm} = 101.325 \text{ kPa}$$

$$1 \text{ l} = 1 \text{ dm}^3$$

$$1 \text{ Pa} = 1 \text{ N m}^{-2}$$

9. At relatively low pressures, it is possible to ignore the constant b in the van der Waals equation to give the equation of state

$$pV^2 - RTV + a = 0$$

Solve this quadratic equation in V using the formula.

10. The position of equilibrium in a system consisting of 0.300 mol of gaseous N_2O_4 and 0.500 mol of gaseous NO_2 can be specified by the extent of reaction x. It can be shown that the value of x is given by the equation

$$4.0730x^2 + 2.0365x + 0.2325 = 0$$

and that x is between -0.350 and -0.300. Use an iterative method to determine the value of x.

Solution chemistry

4

In some respects, the physical chemistry of liquid solutions is more complicated than that of either gases or solids. Molecules in the gas phase have relatively large separations, so the interactions between them are reduced, and
many approximations may be used to simplify the mathematics involved. Conversely, if a solid is crystalline, it can be considered to be composed of many
repeats of a simple unit, which again reduces the complexity of the mathematics required.

In this chapter, we will be concerned only with aqueous solutions, as these
form by far the largest group of interest. Much of what we do will come down to
exploring the forms of the various equations so that they can be of use to us.

4.1 Multiplication in the calculation of activity

We met the concept of activity briefly in Chapter 3, Section 3.9, on page 102.
For a solution of concentration c, the activity a is defined as

$$a = \left(\frac{c}{c^o} \right) \gamma_\pm$$

where c^o is the standard concentration of exactly 1 mol dm^{-3}, and $\gamma_\pm$ is a constant known as the activity coefficient. In dilute solutions, $\gamma_\pm$ is close to 1, so
activities are often taken as being equal to concentrations.

Worked example 4.1

The activity coefficient of a 0.050 mol dm^{-3} solution of NaCl in water is
0.820. Determine the activity of this solution.

Chemical background

Activity coefficients can be determined using a variety of techniques
that include vapour pressure measurements, freezing point depression,
solubility, and electrochemistry. The quantity $\gamma_\pm$ is strictly the mean
activity coefficient; this is related to the activity coefficients γ_+ and γ_-
for positive and negative ions, respectively, by

$$\gamma_\pm = \sqrt{\gamma_+ \gamma_-}$$

Solution to worked example 4.1

We have all the values we require to substitute into the defining equation to give

$$a = \left(\frac{0.050 \, \text{mol dm}^{-3}}{1 \, \text{mol dm}^{-3}} \right) \times 0.820$$

The units cancel top and bottom to leave

$$a = \frac{0.05}{1} \times 0.820 = 0.041$$

Notice that we give the answer to 2 significant figures, as this is the minimum level of precision in the quantity 0.05 mol dm^{-3} in the original data.

4.2 Molality

We are familiar with solutions whose concentration is expressed in units of mol dm^{-3}. Concentration is accordingly defined as the amount of solute in 1 dm^3 of solution. An alternative way of expressing the strength of solution is by using its molality. This has units of mol kg^{-1}, and is defined as the amount of solute in 1 kg of solvent.

Note that the volume of a solution changes with temperature, but the mass of solvent does not. Molality is thus independent of temperature, whereas concentration is not.

4.2.1 *Proportion*

Proportion was introduced in Chapter 3, Section 3.7.3, on page 88. We saw there that two quantities were directly proportional if y/x is constant, and inversely proportional if xy is constant.

Worked example 4.2

0.50 mol of HCl was dissolved in 0.75 kg of water. What amount of HCl should be dissolved in 2.50 kg of water to give a solution of the same molality?

Chemical background

Hydrogen chloride gas consists of covalently bonded HCl molecules with weak interactions between them. These readily dissolve in water to form separate H$^+$ and Cl$^-$; it is the former that gives the resulting solution its acidic properties.

Solution to worked example 4.2

A little thought should suggest that for a series of solutions having the same molality, the amount of HCl will be directly proportional to the mass of water. Consequently, for the first solution we have

$$\frac{n_1}{M_1} = \frac{0.50\,\text{mol}}{0.75\,\text{kg}} = 0.67\,\text{mol kg}^{-1}$$

where n_1 is the amount of HCl and M_1 the mass of water in the first solution.

In this problem, this constant is our constant of proportionality, but you may also realize from the units that it is in fact the molality of the solution. For the second solution we can therefore write

$$\frac{n_2}{2.50\,\text{kg}} = 0.67\,\text{mol kg}^{-1}$$

where n_2 is the amount of HCl in this second solution. If we multiply both sides of this equation by 2.50 kg, we obtain

$$n_2 = 0.67\,\text{mol kg}^{-1} \times 2.50\,\text{kg} = 1.68\,\text{mol}$$

4.3 Raoult's law

Raoult's law allows us to calculate the vapour pressure above a solution that contains a mixture of two liquids. If this vapour pressure is p, then

$$p = x_1 p_1^* + x_2 p_2^*$$

where x_1 and x_2 are the mole fractions of components 1 and 2, respectively, in the mixture, and p_1^* and p_2^* are the vapour pressures of the pure components 1 and 2, respectively. This relationship is illustrated graphically in Figure 4.1.

4.3.1 Straight-line graphs

We saw in Chapter 3, Section 3.6.2, on page 78 that the equation of a straight line is given by

$$y = mx + c$$

where x and y are the variables, m is the gradient, and c is the intercept, as shown in Figure 4.2.

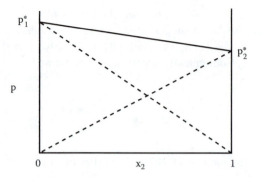

Figure 4.1   Graph of pressure against mole fraction illustrating Raoult's law.

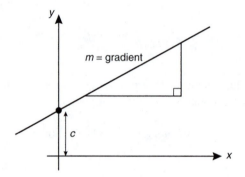

Figure 4.2   The straight-line graph.

Worked example 4.3

Show that Raoult's law can be expressed as a straight line graph of p against the mole fraction x of one of the components.

Chemical background

Raoult's law is observed most closely when the two components in the solution are quite similar, such as benzene and toluene, for example. This is because, in such a solution, the interactions between benzene and benzene, toluene and toluene, and benzene and toluene will be similar.

Solution to worked example 4.3

The important conceptual step in this problem is the realization that

$$x_1 + x_2 = 1$$

We can rearrange this expression by subtracting x_2 from either side to give

$$x_1 + x_2 - x_2 = 1 - x_2$$

or

$$x_1 = 1 - x_2$$

We can now substitute this into the defining equation for Raoult's law to give

$$p = x_1 p_1^* + (1 - x_1)p_2^*$$

Removing the brackets (as in Chapter 3, Section 3.4.1, on page 59) gives us

$$p = x_1 p_1^* + p_2^* + x_1 p_2^*$$

The order of addition is unimportant, so we can write this as

$$p = x_1 p_1^* + x_1 p_2^* + p_2^*$$

Collecting the terms in x_1 now leads to

$$p = (p_1^* + p_2^*)x_1 + p_2^*$$

which is actually the equation of a straight line with variables x_1 and p, gradient $(p_1^* + p_2^*)$, and intercept p_2^*.

Notice that if we did take measurements of x_1 and p, p_2^* is given directly by the intercept, and once we have its value, we can find that of p_1^* from the gradient.

Finally, notice that this problem could also be worked through with x_2 as the x-axis variable rather than x_1.

4.3.2 Proportion

Proportion was introduced in Chapter 3, Section 3.7.3, on page 88. We saw there that two quantities were

- directly proportional if y/x is constant
- inversely proportional if xy is constant

Worked example 4.4

Raoult's law can also be written as

$$p = p_1 + p_2$$

where p_1 and p_2 denote the contributions to the overall pressure of components 1 and 2. These are known as partial pressures and are given by

$$p_1 = x_1 p_1^* \text{ and } p_2 = x_2 p_2^*$$

using the same terms as before.

Air at a temperature of 298 K and a pressure of 786 mm of mercury contains nitrogen at a partial pressure of 613 mm of mercury. Use this data to calculate the percentage of nitrogen in the air.

Chemical background

The phrase "mad as a hatter" is believed to come from the tale *Alice in Wonderland*, in which the Mad Hatter had been poisoned by mercury in the hats he made.

In a mercury barometer the height of mercury is proportional to the barometric pressure. This depends on the altitude and on weather conditions.

Solution to worked example 4.4

Dividing the first equation for partial pressure by p_1^* gives

$$x_1 = \frac{p_1}{p_1^*}$$

Now, substituting in values for p_1 and p_1^* gives

$$x_1 = \frac{613 \, \text{mm Hg}}{786 \, \text{mm Hg}} = 0.78$$

Because we were asked for the percentage value, we have to multiply the fraction by 100 to give the familiar value of 78%.

4.4   The Debye–Hückel equation

The Debye–Hückel limiting law expresses the mean activity coefficient $\gamma_\pm$ in terms of the ionic strength I and the charges z_+ and z_- on the positive and negative ions, respectively:

$$\log \gamma_\pm = -0.51 |z_+ z_-| \sqrt{\frac{I}{\text{mol dm}^{-3}}}$$

The symbol "$|x|$" means "take the value of x without regard to sign"; this is called the *modulus of x*. Thus $|-5|$ would be 5, and $|5|$ would also be 5. The symbol "log" is used to refer to logarithms to the base 10, and the reason for dividing I by units of mol dm^{-3} is to ensure that the quantity on the right-hand side of the equation has no units. This must be true on the left-hand side of the equation as the quantity is logarithmic and the logarithm of a number has no units. As we saw in Section 4.1 on page 115, the mean activity coefficient $\gamma_\pm$ is defined as

$$\sqrt{\gamma_+ \gamma_-}$$

for a 1:1 electrolyte, with γ_+ and γ_- being the activity coefficients of the positive and negative ions, respectively.

The Debye–Hückel limiting law gives reasonable results at low concentrations, but the deviations from it as concentrations increase are considerable. The reduction in the activity coefficient of an ion is due to its interactions with the surrounding ionic atmosphere; the effect increases with charge and in solvents of low dielectric constant, as both these factors increase the magnitude of the interactions.

The relative dielectric constant is defined as the ratio of the dielectric constant of the medium to the dielectric constant of a vacuum. Consequently, it has no units. Typical values of the relative dielectric constant are 78.54 for water and 2.015 for cyclohexane at 25°C.

4.4.1 Logarithms

The subject of logarithms was discussed in Chapter 3, Section 3.6.1, on page 75. Here we saw that the term "log" is used to refer to logarithms to base 10, and that if

$$a = 10^c$$

then

$$c = \log_{10} a$$

or more usually

$$c = \log a$$

Worked example 4.5

Use the Debye–Hückel limiting law to calculate the ionic strength of a solution of potassium chloride having a mean activity coefficient of 0.927.

Chemical background

The ionic strength I of a solution is defined by the equation

$$I = 0.5 \sum_i m_i z_i^2$$

where m_i is the molality of the ions i in the solution, and z_i is their charge. In dilute solutions, it is usually more convenient to replace the molalities m_i by concentrations c_i. The expression then becomes

$$I = 0.5 \sum_i c_i z_i^2$$

The rates of ionic reactions are found to vary between solutions of different ionic strength, and the second-order rate constant k is given by the equation

$$\log k = \log k_0 + 1.02 \, z_A \, z_B \sqrt{\frac{I}{\mathrm{mol\,dm^{-3}}}}$$

where k_0 is a constant that depends upon the concentrations of the activated complexes present.

Solution to worked example 4.5

The first stage in this problem is to rearrange the equation that expresses the Debye–Hückel limiting law so that the term in ionic strength appears on its own on the left-hand side. Dividing both sides by $-0.51 \, |z_+ z_-|$ and reversing the equation gives us the expression

$$\sqrt{\frac{I}{\text{mol dm}^{-3}}} = \frac{\log \gamma_\pm}{-0.51 |z_+ z_-|}$$

In this case, it is probably easier to obtain a value for

$$\sqrt{\frac{I}{\text{mol dm}^{-3}}}$$

We evaluate the logarithm using the 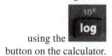 button on the calculator.

and then square it, rather than trying to calculate I directly. Because, in solution, potassium chloride consists of K^+ and Cl^- ions, the values of z_+ and z_- are 1 and -1, respectively. Substituting in the rearranged equation just presented leads to

$$\sqrt{\frac{I}{\text{mol dm}^{-3}}} = \frac{\log 0.927}{-0.51 \times |1 \times (-1)|}$$

Because $|1 \times (-1)| = |-1| = 1$, we then have

$$\sqrt{\frac{I}{\text{mol dm}^{-3}}} = \frac{-\log 0.927}{0.51}$$

$$= \frac{-(-0.0329)}{0.51}$$

$$= 0.0645$$

Use the x^2 button to square this quantity.

(Note that the value of $\log 0.927$ is negative.) To obtain the ionic strength, we now have to square both sides to give

$$\frac{I}{\text{mol dm}^{-3}} = 0.0645^2 = 0.0042$$

and therefore $I = 0.0042$ mol dm^{-3}.

The molar conductivities of K^+ and Cl^- are 73.5 and 76.4 S cm^2 mol^{-1}, respectively, at a concentration of 0.01 mol dm^{-3}. In practice, the conductivity varies with concentration.

4.5   Ostwald's dilution law

The molar conductivity Λ of an ionic solution is a measure of how well it conducts electricity, and is measured in units of Ω^{-1} cm^2 mol^{-1} or S cm^2 mol^{-1}. (Ω

is the symbol for the ohm, the unit of electrical resistance, and S is the symbol for siemens where $1\ S = 1\ \Omega^{-1}$.) It is possible to extrapolate measured values of Λ to zero concentration and obtain a value for Λ_0, the molar conductivity at infinite dilution.

For the dissociation of an electrolyte XY according to the equation

$$XY = X^+ + Y^-$$

the equilibrium constant K_c, in terms of concentration, is defined as

$$K_c = \frac{[X^+][Y^-]}{[XY]}$$

but it is also possible to define the dimensionless equilibrium constant K in terms of the molar conductivities:

$$K = \frac{c\left(\dfrac{\Lambda}{\Lambda_0}\right)^2}{1 - \left(\dfrac{\Lambda}{\Lambda_0}\right)}$$

which is a satisfactory expression for many weak electrolytes.

4.5.1   Discontinuities

A function has a *discontinuity* when it is not defined for every possible input value. For example, consider the function defined by

$$f(x) = \frac{x+1}{x-4}$$

For a function to be defined, there has to be a single value that results from evaluating the function for a given value of x. If we evaluate the function f for the first few integers, we obtain

$$f(0) = \frac{1}{-4} = -\frac{1}{4}$$

$$f(1) = \frac{2}{-3} = -\frac{2}{3}$$

$$f(2) = \frac{3}{-2} = -\frac{3}{2}$$

$$f(3) = \frac{4}{-1} = -4$$

$$f(4) = \frac{5}{0}$$

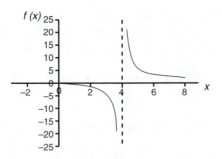

Figure 4.3   Graph of $f(x) = \dfrac{x+1}{x-4}$.

When $x = 4$, the denominator of $f(x)$ is zero. Division by zero results in an infinite number of possible answers, none of which is acceptable, so $f(x)$ is not defined when $x = 4$. A plot of the function $f(x)$ is shown in Figure 4.3. Notice that if the value of x is very close to 4, then the function is still defined; it is discontinuous only at the exact value $x = 4$.

Worked example 4.6

The expression of Ostwald's dilution law just given is an example of a function that has a discontinuity. In this case, it is not possible to obtain a value of K for every value of Λ. For what value or values of Λ is K undefined?

Chemical background

For an electrolyte MX, the degree of dissociation α is defined as

$$\alpha = \frac{m_{x^-}}{m}$$

where m_{x^-} is the molality of the dissociated anion, and m is the overall molality of the MX solution.

For strong electrolytes, the concept of degree of dissociation does not apply. If they are virtually fully ionized, the degree of dissociation α will be equal to 1, and the properties of their solutions will depend to a large extent on the nature of the solvent.

Solution to worked example 4.6

In the equation

$$K = \frac{c\left(\dfrac{\Lambda}{\Lambda_0}\right)^2}{1-\left(\dfrac{\Lambda}{\Lambda_0}\right)}$$

K will be undefined when the denominator is zero, i.e., when

$$1 - \frac{\Lambda}{\Lambda_0} = 0$$

$$\frac{\Lambda}{\Lambda_0} = 1$$

It follows that K is not defined when $\Lambda = \Lambda_0$.

4.6   Partial molar volumes

When two solutions are mixed, the total volume of the resulting solution is not necessarily the sum of the individual volumes. In fact, the total volume V of the solution can be expressed by the equation

$$V = n_1 V_1 + n_2 V_2$$

where n_1 and n_2 are the amounts of components 1 and 2, and V_1 and V_2 are the corresponding respective partial molar volumes. The partial molar volume is actually defined as a partial derivative:

$$V_1 = \left(\frac{\partial V}{\partial n_1} \right)_{T,p,n_2,\dots}$$

Typical values of partial molar volumes are 16.98 cm³ mol⁻¹ and 57.60 cm³ mol⁻¹ for water and ethanol, respectively, in a solution containing 0.42 mol fraction of ethanol.

The subscripts T, p, n_2, ... indicate that the temperature, pressure, and amount of component 2 are all held constant as we consider the variation of the volume V with the amount n_1 of component 1. The dots allow for solutions composed of more than two components, and represent n_3, n_4, etc.

4.6.1   Functions

The subject of functions was discussed in Chapter 3, Section 3.4.3, on page 64. The value of a function $f(x)$ is found by substituting a specific value of x into the expression everywhere that the symbol for x appears.

Worked example 4.7

The partial molar volume V_1 of water in a solution of potassium sulfate is given as a function of molality m by the equation

$$\frac{V_1}{\text{cm}^3} = -0.1094 \left(\frac{m}{m^\circ} \right)^{\frac{3}{2}} - 0.0002 \left(\frac{m}{m^\circ} \right)^2 + 17.963$$

where m° is the standard molality of 1 mol kg⁻¹.
   Calculate the partial molar volume of water when $m = 0.10$ mol kg⁻¹.

Chemical background

Although partial molar volumes are defined in terms of the amounts n_1 and n_2 of solvent and solute, for practical purposes we often work in terms of the molality m of the solute. This is defined as the amount of solute per kg of solvent, so if we actually have (or can assume) 1 kg of solvent, then m will be the same as n_2.

The conversion from concentration to molality can be made if we know the density of the solution.

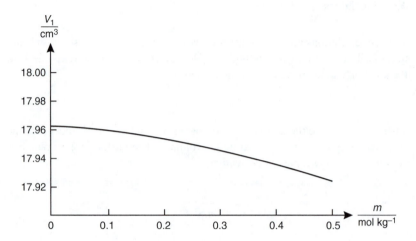

Figure 4.4 Graph of the function describing the partial molar volume of water in a solution of potassium sulfate.

Solution to worked example 3.7

The expression for V_1 can be written using the function notation as

$$\left(\frac{V_1}{cm^3}\right)\left(\frac{m}{m^o}\right) = -0.1094\left(\frac{m}{m^o}\right)^{\frac{3}{2}} - 0.0002\left(\frac{m}{m^o}\right)^2 + 17.963$$

and its graph is shown in Figure 4.4. This expression appears cumbersome at first sight, but inclusion of the molality term on the left of the equals sign does emphasize that this is the variable in the function V_1. We can calculate V_1 for 0.10 mol kg^{-1} by direct substitution into the function expression. This gives us

$$\left(\frac{V_1}{cm^3}\right)\left(\frac{0.10\,mol\,kg^{-1}}{1\,mol\,kg^{-1}}\right) = -0.1094\left(\frac{0.10\,mol\,kg^{-1}}{1\,mol\,kg^{-1}}\right)^{\frac{3}{2}}$$

$$-0.0002\left(\frac{0.10\,mol\,kg^{-1}}{1\,mol\,kg^{-1}}\right)^2 + 17.963$$

and subsequently

$$\left(\frac{V_1}{cm^3}\right)(0.1) = -0.1094 \times 0.10^{\frac{3}{2}} - 0.0002 \times 0.10^2 + 17.963$$

Powers are evaluated using

the ⌃ button.

Evaluating this expression term by term using a calculator gives us

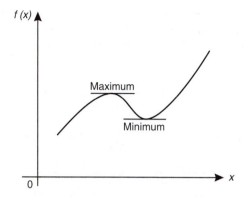

Figure 4.5   Graph of a function $f(x)$ that has both a maximum and a minimum.

$$\frac{V_1(0.1)}{cm^3} = -0.1094 \times 0.03162 - 2 \times 10^{-6} + 17.963$$

$$= -0.00346 - (2 \times 10^{-6}) + 17.963$$

The second term of 2×10^{-6} is much smaller than the other two and can be safely neglected because the final term is given to three decimal places. We are now left with

$$\frac{V_1(0.10)}{cm^3} = -0.00346 + 17.963$$

$$= 17.95954$$

We are therefore restricted to three decimal places in our answer, which will be 17.960 cm^3.

4.6.2   Stationary points

We saw in Chapter 3, Section 3.5.1, on page 66 that we could think of the derivative $df(x)/dx$ as being the gradient of the graph of $f(x)$ against x, which could also be found by drawing the tangent to the curve at the point of interest. Suppose we have a function $f(x)$ whose graph is shown in Figure 4.5. This has a maximum (peak) and a minimum (trough) at the points indicated; these are known as *stationary points*. Notice that at both the maximum and the minimum the tangent to the curve is horizontal, and consequently its gradient has a value of zero.

This fact gives us a way of determining the points at which a function $f(x)$ has a stationary point. If the tangent to the curve is zero, then

$$\frac{df}{dx} = 0$$

It is also possible to identify the nature of the stationary points found. This is done by looking at the value of the second derivative d^2f/dx^2. As its name implies, the second derivative is found by differentiating df/dx again. In symbols this is expressed as

$$\frac{d^2f}{dx^2} = \frac{d}{dx}\left(\frac{df}{dx}\right).$$

The relationship between d^2f/dx^2 and the type of stationary point is as follows:

Value of $\dfrac{d^2f}{dx^2}$	Type of stationary point
Negative	Maximum
Zero	Point of inflexion
Positive	Minimum

A point of inflexion, as in the middle of the letter S, is both a maximum and minimum stationary point.

If

$$f(x) = x^2 - 3x$$

then

$$\frac{df(x)}{dx} = 2x - 3$$

It follows that $\dfrac{df(x)}{dx} = 0$ when $2x - 3 = 0$
so that $2x = 3$ or $x = 3/2$.

Returning to our expression for $\dfrac{df(x)}{dx}$ we have

$$\frac{d^2f(x)}{dx} = \frac{d}{dx}\left(\frac{df(x)}{dx}\right) = \frac{d}{dx}(2x - 3) = 2$$

Because this value is positive, it indicates that the stationary point at $x = 3/2$ is a minimum.

Worked example 4.8

The total volume V of a solution of magnesium sulfate is given in terms of its molality m as

$$\frac{V}{\text{cm}^3} = 1001.38 - 4.86\left(\frac{m}{m^\circ}\right) + 34.69\left(\frac{m}{m^\circ}\right)^2$$

where m° is the standard molality of 1 mol kg^{-1}. For what value of the molality will the volume be at a minimum?

Chemical background

Logically, we might expect the volume to be at a minimum when no magnesium sulfate is present and then to increase steadily as more salt is added until the solution becomes saturated. In fact, the partial molar volume of magnesium sulfate extrapolated to zero concentration has a negative value. The hydration of the Mg^{2+} and SO_4^{2-} ions causes the structure of water to break down and collapse.

Solution to worked example 4.8

To obtain the value of m when V will be a minimum, we need to solve the equation

$$\frac{dV}{dm} = 0$$

To differentiate the expression for V, we need to recall the standard rules

$$\frac{d}{dx}(x) = 1$$

$$\frac{d}{dx}(x^2) = 2x$$

The derivative is then given by

$$\frac{dV}{dm} = 0 - (4.86 \times 1) + (34.69 \times 2m)$$

$$= -4.86 + 69.38m$$

Setting dV/dm to equal zero will allow us to find the value of m at the stationary point:

$$-4.86 + 69.38m = 0$$

To solve this equation we add 4.86 to both sides to give

$$69.38m = 4.86$$

and then divide both sides by 69.38:

$$m = \frac{4.86}{69.38} = 0.070 \, \text{mol kg}^{-1}$$

It would be easy to think that this was the final answer to the question; however, we have not yet shown whether this value gives a maximum or a minimum value of V. Remember that we can do this by looking at the value of d^2V/dm^2, which we obtain by differentiating our expression for dV/dm once more:

$$\frac{d^2V}{dm^2} = \frac{d}{dm}\left(\frac{dV}{dm}\right)$$

$$= \frac{d}{dm}(-4.86 + 69.38m)$$

$$= 0 + (69.38 \times 1)$$

$$= 69.38$$

Again, no substitution of the value of m is required. Because 69.38 is a positive value, $m = 0.070$ mol kg^{-1} gives V its minimum value.

Exercises

1. If p is defined as

$$p = 3 - 2 \ln x$$

where $x = y/z$, evaluate p when $y = 7.5$ and $z = 2.8$.

2. If the value of x^2y is constant, how are x and y related?

3. The variables x and y take the following values

x	−2	−1	0	1	2
y	12	3	0	3	12

Deduce the relationship between x and y.

4. If r is directly proportional to both x and y, and inversely proportional to z^2, obtain a general equation that relates r to x, y, and z.

5. Solve the equation

$$\log_x 81 = 3$$

for x.

6. Solve the equation

$$\log_2 x = 4$$

7. For the following equations, state which plot would give a straight line, and what the resulting values of the gradient and intercept would be.

(a) $y = 4 + 6 \ln x$

(b) $\log y = 2\sqrt{x}$

(c) $\ln y = 2 \log x - 3$

8. For the following equations, state which plot would give a straight line, and what the resulting values of the gradient and intercept would be.

(a) $t = s^2 - 5$

(b) $3w = 9v + 27$

(c) $a - b = 14$

9. For the following equations, state which plot would give a straight line, and what the resulting values of the gradient and intercept would be.

(a) $y = \dfrac{4}{x} + 7$

(b) $\log 2y = \dfrac{1}{x^2} + 6$

(c) $\dfrac{1}{y^2} = \dfrac{1}{x} + \dfrac{1}{2}$

10. Two variables x and y take the following values:

x	−1.5	−1.0	−0.5	0.0	0.5	1.0	1.5	2.0
y	−3.1	−2.2	−0.9	−0.4	0.5	1.4	2.3	3.2

By plotting this data, determine the relationship between x and y.

11. Two variables x and y are related by an equation of the form

$$y = ax^3 + b$$

Use the data below to determine the values of a and b:

x	−3	−2	−1	0	1	2	3
y	−117.2	−37.4	−8.0	−3.8	0.4	29.8	109.6

12. At what values of x do the following functions of x have discontinuities?

(a) $f(x) = \dfrac{x}{x^2 - 9}$

(b) $f(x) = \dfrac{3x + 1}{2x + 3}$

13. At what values of x and/or y do the following functions of x and y have discontinuities?

(a) $f(x, y) = \dfrac{x + y + 1}{x^2 + y^2}$

(b) $f(x, y) = \dfrac{2y}{x(x-1)}$

14. The function $h(x, y)$ is defined as

$$h(x, y) = \frac{x(y-1)}{y(x-1)}$$

and has discontinuities along two lines. Give the value of x that defines one line, and the value of y that defines the other line.

15. For the function defined by

$$f(x) = 9x^5 + 3x^3 - 2x + 1$$

determine $\dfrac{df(x)}{dx}$ and $\dfrac{d^2 f(x)}{dx^2}$.

16. For the function defined by

$$g(y) = 2y^2 - \frac{3}{y}$$

determine $\dfrac{d^2 g(y)}{dy^2}$.

17. Locate the stationary points of the function

$$f(x) = 2x^3 - 24x + 7$$

18. Locate and identify the stationary points of the function

$$g(x) = 3x^2 - 2x + 1$$

19. Locate the stationary points of the function

$$h(x) = x^2 - \sqrt{x}$$

20. Find the stationary points of

$$f(x) = x^3 - 3x^2 + 3x - 1$$

Problems

1. A solution of potassium chloride in water has a concentration of 0.272 mol dm^{-3} and a density of 1.012 g cm^{-3}. Calculate the molality of the solution.

2. For each of the following equations, state which quantities you would plot to give a straight line, and the values of the resulting gradient and intercept:

(a) $\pi = cRT$

Here π is the osmotic pressure that varies according to concentration; c, R is the gas constant; and T is the specified absolute temperature.

(b) $\log k = \log k_0 + 1.02 z_A z_B \sqrt{I}$

Here k is the rate constant for a reaction, which varies according to the ionic strength I. The rate constant for the limit of unit activity coefficients is k_0, and z_A and z_B are the charges on ions A and B, respectively.

(c) $\Lambda = \Lambda_0 - (P + Q\Lambda_0)\sqrt{c}$

Here Λ is the molar conductivity of an electrolyte solution of concentration c, Λ_0 is the molar conductivity at infinite dilution, and P and Q are constants.

3. The Nernst equation gives the electromotive force E in terms of its value E° under standard conditions, the reaction quotient Q, the absolute temperature T, and the numerical amount of charge transferred, z, as

$$E = E^\circ - \left(\frac{RT}{zF}\right) \ln Q$$

where R is the gas constant and F is the Faraday constant.

   In a series of measurements made on an electrochemical cell, the cell potential E was determined as a function of (a) reaction quotient Q and (b) temperature T. What information can be deduced from an appropriate straight line graph in each case?

4. A volume 25 cm^3 of a solution of sodium chloride of concentration 1.0 mol dm^{-3} was mixed with 75 cm^3 of a solution of sodium chloride of concentration 0.1 mol dm^{-3}. What is the concentration of the resulting solution?

5. Fick's first law relates the diffusive flux J to the concentration gradient by the equation

$$J = -DA\frac{\partial c}{\partial x}$$

where D is the diffusion coefficient, A the area passed by the solute, c the concentration, and x the distance.

   The net rate of increase of concentration with time is given by the expression

$$\frac{\partial c}{\partial t} = -\frac{1}{A}\frac{\partial J}{\partial x}$$

Use Fick's first law to eliminate J from this equation, assuming that D is independent of x.

6. (a) The molar mass M of a solute is given in terms of the sedimentation coefficient s and the diffusion coefficient D:

$$M = \frac{RTs}{D(1 - V_1\rho)}$$

where R is the gas constant, T the absolute temperature, and V_1 the volume per unit mass of the solute. When will it *not* be possible to calculate M using this equation?

(b) For a solute that undergoes a dimerization reaction with equilibrium constant K, it can be shown that

$$K = \frac{K_b(K_b m - \Delta T_b)}{(2\Delta T_b - K_b m)^2}$$

where ΔT_b is the boiling point elevation, K_b the ebullioscopic constant, and m the molality. Under what condition is K undefined according to this equation?

7. (a) If v is the total number of ions resulting from the complete dissociation of a molecule, and i is the van't Hoff factor related to the colligative properties of electrolytes, then the degree of dissociation α is given by the equation

$$\alpha = \frac{i - 1}{v - 1}$$

By considering possible values of v, decide whether α is a continuous function for electrolytes.

(b) For a single 1:1 electrolyte, the dissociation constant K is given by the expression

$$K = \frac{c^2\alpha^2}{1 - \alpha}$$

where c is the concentration and α is the degree of dissociation. Under what circumstances is K not defined?

8. The total volume V of a sodium chloride solution at 25°C is given by the expression

$$\frac{V}{\text{cm}^3} = 1003.0 + 16.4\left(\frac{n}{\text{mol}}\right) + 2.1\left(\frac{n}{\text{mol}}\right)^{\frac{3}{2}} + 0.003\left(\frac{n}{\text{mol}}\right)^{\frac{5}{2}}$$

where n is the amount of salt in 1 kg of water. Calculate the partial molar volume V_{NaCl}, which is equal to the derivative dV/dn when $n = 0.15$ mol kg^{-1}.

9. The volume V of a solution of an alcohol in water is given by the equation

$$\frac{V}{cm^3} = 1004.08 + 3.9062\left(\frac{m}{m^\circ}\right) - 2.4362\left(\frac{m}{m^\circ}\right)^2$$

here m is the molality and $m^\circ = 1$ mol kg^{-1}. Determine the maximum volume of this solution.

10. The density $\rho(x)$ of a solution of a certain alcohol in water is described by the function

$$\frac{\rho(x)}{g\,cm^{-3}} = 0.982 - 0.265x + 0.306x^2$$

at a temperature of 25°C, with x being the mole fraction of alcohol in the water. Determine the value of the mole fraction for which the solution density is a maximum or a minimum. Identify the nature of the stationary point.

Kinetics 5

In contrast to the thermodynamics discussed in Chapter 3, the range of mathematical techniques required for a study of kinetics is somewhat smaller. We will make much use of a technique called integration, which is essentially the reverse of the differentiation process we have already met. Analyzing the rates of reactions normally involves taking a derivative, which describes the rate of change of concentration with time, and then reversing the differentiation process to obtain the original function.

5.1   Using a rate equation

For a chemical reaction between molecules A and B, the rate v of the reaction will be given by an equation of the form

$$v = k[A]^x[B]^y$$

where k is a constant known as the rate constant, which varies with temperature. [A] and [B] are the concentrations of A and B, respectively, raised to powers x and y, which are often, but not always, integers. The reaction is said to have order x with respect to A, order y with respect to B, and overall order $x + y$.

Rate always has units of mol dm^{-3} s^{-1}; because x and y take different values for different reactions, it follows that k will also have different units for different reactions.

Worked example 5.1

For the reaction

$$2\,NO_{(g)} + Cl_{2(g)} \rightarrow 2\,NOCl_{(g)}$$

the rate v is given by the equation

$$v = k[NO]^2[Cl_2]$$

where k is the rate constant whose value is 3.84×10^{-4} dm^6 mol^{-2} s^{-1}.

(a) Calculate the rate when a vessel of volume 2.50 dm^3 contains 4.00 mol of NO and 3.50 mol Cl$_2$.
(b) What is the effect on the rate if the amount of NO is halved?

Chemical background

Nitrosyl chloride, NOCl, can be found as an orange gas, and is both toxic and corrosive. Its oxidizing properties lead to it being used as a bleaching agent in flour, and it can also be found in detergents.

Solution to worked example 5.1

(a) We can calculate the concentrations of NO and Cl_2 by dividing the amount present by the volume:

$$[NO] = \frac{4.00\,mol}{2.50\,dm^3} = 1.60\,mol\,dm^{-3} \quad [Cl_2] = \frac{3.50\,mol}{2.50\,dm^3} = 1.40\,mol\,dm^{-3}$$

We can now substitute directly into the equation given:

$$v = (3.84 \times 10^{-4}\,dm^6\,mol^{-2}\,s^{-1}) \times (1.60\,mol\,dm^{-3})^2 \times (1.40\,mol\,dm^{-3})$$

$$= (3.84 \times 10^{-4}\,dm^6\,mol^{-2}\,s^{-1}) \times (2.56\,mol^2\,dm^{-6}) \times (1.40\,mol\,dm^{-3})$$

$$= 1.38 \times 10^{-3}\,mol^{(-2+2+1)}\,dm^{(6-6-3)}\,s^{(-1)}$$

$$= 1.38 \times 10^{-3}\,mol\,dm^{-3}\,s^{-1}$$

because, as the units of each quantity are multiplied, we add the indices each time.

(b) Because the [NO] term in the equation is squared, if we halve this, the rate will change by a factor of $(½)^2$ or ¼; in other words, it is reduced by a factor of 4.

5.2   Rates of change

We saw in Chapter 3, Section 3.5.1, on page 66 that we can consider both average and instantaneous rates of change. For chemical reactions, we normally monitor the change in concentration, or a related quantity, of the reactants or products with time. If we know the initial concentration and its value after a specified time interval, it is possible to calculate the average rate of reaction.

Worked example 5.2

Propionaldehyde is prepared from ethane, hydrogen, and carbon monoxide by means of an homogeneously catalyzed hydroformylation reaction.

In the reaction between propionaldehyde and hydrocyanic acid, the concentration of the acid was 0.0902 mol dm^{-3} after 5.2 min and 0.1652 mol dm^{-3} after 20.0 min. What is the average rate of reaction?

Chemical background

Full details of this study are given in the *Journal of the American Chemical Society* (**75**, 3106, 1953). The effect of variations in the ionic strength, pH, and buffer solution on the reaction rate were monitored in acetate buffers. The reaction was found to be second order.

Solution to worked example 5.2

We need to divide the change in concentration by the change in time, remembering that the change is calculated as final value minus initial value for both quantities:

$$\text{change in concentration} = (0.1652 - 0.0902) \text{ mol dm}^{-3}$$
$$= 0.0750 \text{ mol dm}^{-3}$$

$$\text{change in time} = (20.0 - 5.2) \text{ min}$$
$$= 14.8 \text{ min}$$

So,

$$\text{average rate of reaction} = \frac{0.0750 \text{ mol dm}^{-3}}{14.8 \text{ min}}$$

$$= 5.07 \times 10^{-3} \text{ mol dm}^{-3} \text{ min}^{-1}$$

This is a perfectly correct way of presenting the answer. However, it is often useful to express such quantities in terms of SI units. In this case, the SI unit of time is seconds (s), and 60 s = 1 min. Substituting this information into our answer gives

$$\text{average rate of reaction} = 5.07 \times 10^{-3} \text{ mol dm}^{-3} (60 \text{ s})^{-1}.$$

If we apply the power of −1 to both the number and unit in the bracket, we get

$$\text{average rate of reaction} = 5.07 \times 10^{-3} \text{ mol dm}^{-3} \times 60^{-1} \text{ s}^{-1}$$

Now, a power of −1 is equivalent to calculating a reciprocal ("one over a quantity")

$$60^{-1} = \frac{1}{60}$$

and so

$$\text{average rate of reaction} = \frac{5.07 \times 10^{-3} \text{ mol dm}^{-3} \text{s}^{-1}}{60}$$

$$= 8.45 \times 10^{-5} \text{ mol dm}^{-3} \text{ s}^{-1}$$

Worked example 5.3

In the reaction

$$H_{2(g)} + Br_{2(g)} \rightarrow 2HBr_{(g)}$$

the initial bromine concentration of 0.321 mol dm^{-3} falls to a value of 0.319 mol dm^{-3} after 0.005 s. Assuming that this time interval is small compared to the time taken for the reaction to reach completion, determine expressions for the rate of change of concentration of each chemical species with respect to time.

Chemical background

Once a complex reaction is under way, it is possible for subsidiary reactions to occur between the products and reactants. Because of this, the system may only be completely defined in terms of the species present and their states at the start of the reaction, and measurements of rate will then be more reliable than later on. It is, of course, necessary to be able to make measurements on a timescale that is short relative to that of the overall reaction.

Solution to worked example 5.3

We saw in Chapter 3, Section 3.5.1, on page 66 that the value of $\dfrac{df(x)}{dx}$ could be considered to be the gradient of the line joining two points that had very similar values of x. Because we are told that the time interval of 0.005 s is very short in this case, the rate of change of bromine can be taken to be its instantaneous value and can therefore be represented by the derivative

$$\frac{d[\text{Br}_2]}{dt}$$

The square brackets are used to denote concentration, and the subscript denoting the state is omitted for clarity.

To obtain the rate of change, we need to divide the increase in concentration by the increase in time, remembering that we always subtract initial values from final.

$$\text{change in concentration} = (0.319 - 0.321)\ \text{mol dm}^{-3}$$

$$= -0.002\ \text{mol dm}^{-3}$$

Because we have defined the change in concentration as an increase, the negative value indicates that the concentration actually decreases, as we can see in the question. Now,

$$\text{change in time} = 0.005\ \text{s}$$

so the instantaneous rate of change is given by the expression

$$\frac{d[\text{Br}_2]}{dt} = \frac{-0.002\ \text{mol dm}^{-3}}{0.005\ \text{s}}$$

$$= -0.4\ \text{mol dm}^{-3}\ \text{s}^{-1}$$

We also need to obtain expressions for the rates of change of H_2 and HBr. Because hydrogen and bromine react in a 1:1 ratio, they are being consumed at the same rate, so we can write

$$\frac{d[H_2]}{dt} = -0.4 \, \text{mol dm}^{-3} \, \text{s}^{-1}$$

The equation also shows us that for every molecule of bromine consumed, two molecules of hydrogen bromide are produced. Therefore, although the concentration of bromine decreases, that of hydrogen bromide increases. Because $\frac{d[Br_2]}{dt}$ has a negative value, it follows that $\frac{d[HBr]}{dt}$ will have a positive value that is twice the magnitude:

$$\frac{d[HBr]}{dt} = -2\frac{d[Br_2]}{dt}$$

$$= -2 \times (-0.4 \, \text{mol dm}^{-3} \, \text{s}^{-1})$$

$$= 0.8 \, \text{mol dm}^{-3} \, \text{s}^{-1}$$

5.3   Zero-order reactions

From our definition in Section 5.1 on page 137, a zero-order reaction is one for which the sum of the powers in the rate equation is zero. Taking our general expression

$$v = k[A]^x[B]^y$$

we need to have $x + y = 0$ for a zero-order reaction. This can most obviously be obtained when $x = y = 0$. Then

$$v = k[A]^0[B]^0$$

and because, as we saw in Chapter 1, Section 1.5.3, on page 15, any number raised to a zero power is one, the expression for the rate is therefore

$$v = k \times 1 \times 1 = k$$

i.e., the rate remains constant throughout the reaction and does not depend on the concentrations of the reactants.

5.3.1   Integration

Integration is essentially the reverse of differentiation, so if we take a function, differentiate it, and then integrate it, we should get back to our starting function.

Let us consider what happens if we differentiate the function $f(x) = x^2$. Remember that when we differentiate, we multiply by the power and reduce the power by one, so that we obtain the derivative

$$\frac{df(x)}{dx} = 2 \times x^{2-1} = 2x$$

In words, the derivative of x^2 is $2x$. Because integration is the reverse of this, we can also state that

the integral of $2x$ is x^2

Because the 2 in this expression is a constant, we can divide our statement by two so that it becomes:

the integral of x is $x^2/2$

Such statements in words are cumbersome, particularly when we have more complicated expressions, and so, in mathematical terminology, we write

$$\int x\, dx = \frac{x^2}{2}$$

The precise words to express this are: "the integral of x with respect to x is equal to x squared divided by 2."

Using similar arguments, and remembering that

$$\frac{d(x^3)}{dx} = 3x^2$$

we obtain the result

$$\int x^2\, dx = \frac{x^3}{3}$$

You should begin to see a pattern emerging. In fact, to integrate any polynomial type function, we increase the power by one and divide by the new power. This works for positive and negative powers but not for x^{-1} (alternatively written as $1/x$) because this would involve dividing by zero. In mathematical terminology

$$\int x^n = \frac{x^{n+1}}{n+1} \quad \text{for} \quad n \neq -1$$

There is one complication in considering integration to be the reverse process of differentiation. Applying the standard methods of differentiation gives us the derivatives

$$\frac{d}{dx}(x^3 + 1) = 3x^2$$

$$\frac{d}{dx}(x^3 + 2) = 3x^2$$

because the derivative of any constant term (1 or 2 in these examples) is zero. If we now try to reverse the process by integrating, we obtain

$$\int 3x^2 = \frac{3x^3}{3} = x^3$$

and we obtain no information about the constant term. It is usual, when calculating such *indefinite integrals*, to include a constant of integration, which is usually given by the symbol C. We then obtain

$$\int 3x^2 dx = x^3 + C$$

C can often be calculated if we know the value of the function $f(x)$ at a certain value of x. These known values are substituted into the equation, which can then be solved for C.

More commonly in chemistry, we evaluate what is called the *definite integral*. We are frequently interested in a range of values that we include as limits, and these avoid the need to calculate a constant of integration. For example, if we were asked to calculate the integral of x^3 between $x = 1$ and $x = 2$, we could express this as

$$\int_1^2 x^3 dx$$

where the lower and upper limits are included on the integration sign. To integrate, we raise the power of the function x^3 by one and divide by the new power, to obtain $x^4/4$. This expression is enclosed in square brackets with the limits outside, and so we write

$$\int_1^2 x^3 dx = \left[\frac{x^4}{4} + C \right]_1^2$$

where C is the constant of integration.

We then evaluate the expression at the upper limit and subtract from it the value at the lower limit, and obtain

$$\left[\frac{x^4}{4} + C \right]_1^2 = \left(\frac{2^4}{4} + C \right) - \left(\frac{1^4}{4} + C \right)$$

$$= \left(\frac{16}{4} \right) - \left(\frac{1}{4} \right)$$

$$= \frac{15}{4}$$

Tungsten surface

Figure 5.1   Ammonia decomposing on the surface of hot tungsten.

this time using rounded brackets, if necessary. Notice that the constants of integration are the same in each bracket so cancel and we obtain a definite value for the integral.

Worked example 5.4

The rate equation for a zero-order reaction can be expressed as

$$-\frac{dc}{dt} = k$$

where dc/dt is the rate of change of concentration c of a reactant with time t, and k is the rate constant. Obtain an expression for the concentration c at a time t assuming that the initial concentration value is c_0.

Chemical background

Arsine AsH_3 and stibine SbH_3 both have similar properties to phosphine but decompose more readily into their elements.

A number of heterogeneous reactions obey zero-order kinetics. These include the decompositions of phosphine or ammonia on hot tungsten, shown in Figure 5.1, and the decomposition of hydrogen iodide on a hot gold wire. This is the case as long as the gas pressure is not too low.

Solution to worked example 5.4

As the expression given contains a derivative, we need to reverse the differentiation process by integration. The first step is to separate the variables. This is necessary before we can integrate, and involves rearranging the expression so that all terms in the variable c are on one side of the equation and all terms in the variable t are on the other side. In this example, this can be done by multiplying both sides by the term dt to give

$$-dc = k\, dt$$

It is also usual to make the subject (i.e., the left-hand side of the equation) positive, and this can be achieved by multiplying both sides of the equation by -1 to give

$$dc = -kdt$$

We have now separated the variables, and so this expression is now in a form that can be integrated. We write

$$\int dc = \int -kdt$$

using the integral signs to show integration.

Note that because $-k$ is a constant, it can be brought outside the integral sign:

$$\int dc = -k \int dt$$

We now need to assign limits to avoid having to deal with constants of integration. Because the initial concentration is c_0, this will occur when t is zero, and so, our lower limits are $c = c_0$ and $t = 0$. As we require an expression for c after time t, these are set as our upper limits. This may seem strange at first, but this is a useful tool for obtaining expressions of general use. Including these limits gives us the expression

$$\int_{c_0}^{c} dc = -k \int_{0}^{t} dt$$

It may not be obvious at this stage what quantities we are going to integrate. The terms dc and dt are part of the notation of integration that say "with respect to c" and "with respect to t," respectively. Remember, however, that both the quantities c^0 and t^0 have a value of one, so we can quite justifiably include them in the respective integrals:

$$\int_{c_0}^{c} c^0 dc = -k \int_{0}^{t} t^0 dt$$

We can now apply the rule that, to integrate we increase the power by one and divide by the new power. Thus c^0 integrates to c and t^0 integrates to t. You may find it worthwhile remembering the rule

$$\int dx = x + C$$

This type of standard integral should always include the constant of integration for mathematical correctness even if we do not show it when calculating definite integrals.

We now need to evaluate our integrated expression at the specified limits

$$\left[c \right]_{c_0}^{c} = -k \left[t \right]_{0}^{t}$$

We substitute the limits for the quantities in square brackets, and subtract the lower from the upper to obtain

$$c - c_0 = -k(t - 0)$$

$$c - c_0 = -kt$$

Alternatively, this could be multiplied throughout by −1 to give

$$c_0 - c = kt$$

Worked example 5.5

What are the units of the rate constant for a zero-order reaction?

Chemical background

The base hydrolysis of nitropentacyanoferrate(III), $Fe(CN)_5NO_2^{3-}$, in aqueous solution follows zero-order kinetics at a pressure of 0.1 MPa and a temperature of 25°C when the concentration of OH^- is 0.1 mol dm^{-3}, the rate constant being 1.20 × 10^7 mol dm^{-3} s^{-1}.

The rate constant has different units for reactions of different order. In contrast, the actual rate of a reaction will always have the units mol dm^{-3} s^{-1} because it is defined as the rate of change of a concentration with time.

Solution to worked example 5.5

We need to rearrange the equation

$$c_0 - c = kt$$

to give an expression for k. If we divide both sides of the equation by t, we obtain

$$\frac{c_0 - c}{t} = k$$

The units of k are therefore those of

$$\frac{\text{concentration}}{\text{time}}$$

which are

$$\frac{\text{mol dm}^{-3}}{\text{s}}$$

Because division is the same as multiplying by a reciprocal,

$$\frac{1}{s} = s^{-1}$$

and so the SI units of the zero-order rate constant are mol dm^{-3} s^{-1}. This is as expected, because, for a zero-order reaction, the rate of reaction is equal to the rate constant.

Worked example 5.6

The half-life of a reaction $t_{1/2}$ is the time taken for the concentration of reactant to fall to half its initial value. What is the half-life for a zero-order reaction?

Chemical background

Although the definition of a half-life is given in terms of the concentration relative to its initial value, it does not actually matter when the starting concentration is taken. The time interval for a particular concentration to be halved will remain constant regardless of the starting time.

Solution to worked example 5.6

In this problem, we need to rearrange the integrated rate equation

$$c_0 - c = kt$$

to give an expression for t. This is done by dividing both sides by k to give

$$t = \frac{c_0 - c}{k}$$

From the information given in the question, when t is equal to $t_{1/2}$, c will be equal to 1/2 c_0, i.e., half the initial value. Substituting these values gives us

$$t_{\frac{1}{2}} = \frac{c_0 - \frac{1}{2}c_0}{k}$$

$$= \frac{\frac{1}{2}c_0}{k}$$

$$= \frac{c_0}{2k}$$

Worked example 5.7

If the concentration of a reactant in a zeroth-order reaction was monitored as a function of time, what graph would you plot to obtain the value of the rate constant?

Chemical background

Some of the practicalities of monitoring such reactions are illustrated by the catalytic decomposition of hydrogen iodide on the surface of gold, which was reported by Hinshelwood and Prichard in the *Journal of the American Chemical Society* in 1925 (**127**, 1552). There is a difficulty in following the pressure change during this reaction as the hydrogen iodide attacks the mercury in a manometer. This was overcome by using hydrogen gas as a buffer in the manometer because it has no effect on the rate of reaction.

Solution to worked example 5.7

Whenever we are analyzing data, we try to plot graphs that give us straight lines. The way in which we can compare an expression with the general straight line equation $y = mx + c$ was outlined in Chapter 3, Section 3.6.2, on page 78. Here, x and y are the variables, and m and c are both constants representing the slope of the graph and the intercept on the y axis, respectively.

In this case, we need to rearrange the equation obtained in worked example 5.4 into this form. Note that c and t are the variables. If we start with

$$c - c_0 = -kt$$

and add c_0 to both sides, we get

$$c = -kt + c_0$$

This is now in the required form with c and t as variables, and $-k$ and c_0 as constants. If we plot c on the y-axis and t on the x-axis, we will obtain a straight line with gradient $-k$ and intercept c_0, as shown in Figure 5.2.

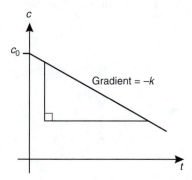

Figure 5.2 Graph of integrated rate equation for a zero-order reaction.

5.4   First-order reactions

The rate equation for a first-order reaction is of the form

$$v = k[A]$$

where [A] is the concentration of a reactant and k is the first-order rate constant.

The rate of reaction now depends on the concentration of reactant, and as this falls during the course of the reaction, so does the rate.

5.4.1   Integration of 1/x

We saw in Section 5.3.1 on page 141 that we could integrate any expression of the form x^n as long as $n \neq -1$. The integration rule involves raising the power of x by one and dividing by the new power but, if $n = -1$, you find you need to divide by zero, and this will not work. In fact, there is another rule for integrating x^{-1} or $1/x$:

$$\int x^{-1} dx = \int \frac{1}{x} dx = \int \frac{dx}{x} = \ln x + C$$

This tells us that when integrating x^{-1} we obtain the natural logarithm $\ln x$. Notice that the three forms of expressing this integral are all equivalent, and you may well meet all of them.

5.4.2   Rules of logarithms

Chapter 3, Section 3.6.1, on page 75 mentioned that, in the past, logarithms were used to assist in the processes of multiplication and division. If we have two numbers X and Y, then

$$\log(XY) = \log X + \log Y$$

$$\log\left(\frac{X}{Y}\right) = \log X - \log Y$$

These rules apply regardless of the base to which the logarithms are taken. They replace the process of multiplication of X by Y by the process of addition of their logs, and the process of division of X by Y by the process of subtraction of log Y from log X, if the logarithms of the required numbers can be obtained.

Notice that if $X = Y$ we have

$$\log (XX) = \log X + \log X = 2 \log X$$

and because

$$XX = X^2$$

it follows that

$$\log X^2 = 2 \log X$$

Also

$$\log\left(\frac{X}{X}\right) = \log X - \log X = 0$$

and because

$$\frac{X}{X} = 1$$

it follows that

$$\log 1 = 0$$

Generally,

$$\log X^n = n \log X$$

for all values of n.

Worked example 5.8

The rate equation for a first-order reaction can be expressed as

$$-\frac{dc}{dt} = kc$$

where dc/dt is the rate of change of concentration c of a reactant with time t, and k is the rate constant. Obtain an expression for the concentration c at a time t assuming that the initial concentration value is c_0.

Chemical background

Many reactions obey first-order kinetics. Examples include

$$2N_2O_{5(g)} \rightarrow 4NO_{2(g)} + O_{2(g)}$$

and

$$SO_2Cl_2 = SO_2 + Cl_2$$

Sulfuryl chloride, which is a colourless fuming liquid, hydrolyzes vigorously to form sulfuric acid and hydrochloric acid. It is used as a chlorinating agent.

Solution to worked example 5.8

Most of the stages in setting up this problem are identical to those in the case of the zero-order reaction in worked example 5.4. To integrate, we must first separate the variables. If the terms containing variable c are to appear on the left-hand side as before, it is necessary to divide both sides by c and multiply both sides by dt before adding the integrals in symbols complete with the initial and final values of c and t. This then results in the expression

$$\int_{c_0}^{c} \frac{dc}{c} = -k \int_{0}^{t} dt$$

As before,

$$\int dt = t$$

although we now know that

$$\int \frac{1}{c} dc = \ln c$$

We then obtain

$$\left[\ln c \right]_{c_0}^{c} = -k \left[t \right]_{0}^{t}$$

which, on substituting the upper and lower limits into the square brackets, gives

$$\ln c - \ln c_0 = -k(t - 0)$$

$$= -kt$$

We saw earlier in this section that

$$\log X - \log Y = \log \left(\frac{X}{Y} \right)$$

so, in this case

$$\ln c - \ln c_0 = \ln \left(\frac{c}{c_0} \right)$$

Putting this into the integrated first-order rate equation gives us the result

$$\ln \left(\frac{c}{c_0} \right) = -kt$$

Worked example 5.9

What are the units of the first-order rate constant?

Chemical background

H_2N—NH_2

An example of a first-order rate constant is the value of $3.3 \times 10^{-4} \text{ s}^{-1}$ for the *cis-trans* isomerization of $Cr(en)_2(OH)_2^+$. The symbol en stands for the ligand ethylenediamine, $NH_2CH_2CH_2NH_2$.

Solution to worked example 5.9

As in the case of worked example 5.5 for a zero-order reaction, we need to rearrange our integrated rate equation to make the rate constant, k, the subject. Dividing both sides of the equation

$$\ln\left(\frac{c}{c_0}\right) = -kt$$

by $-t$ gives

$$\frac{\ln\left(\dfrac{c}{c_0}\right)}{-t} = k$$

Because logarithmic quantities have no units, k will have units equivalent to those of $1/t$ or t^{-1}, i.e., s^{-1}.

Worked example 5.10

What is the half-life of a first-order reaction?

Chemical background

The fact that different expressions emerge for the half-lives of reactions of different orders leads to a very quick way of determining whether a reaction is first order or not. The half-lives of zero- and second-order reactions depend on the initial reactant concentration, whereas the half-life of the first-order reaction does not.

Solution to worked example 5.10

We saw in worked example 5.6 that the half-life $t_{1/2}$ of a reaction is the time taken for the concentration to fall to half its initial value. Rearranging the integrated rate equation for a first-order reaction to give an expression for t gives

$$t = -\frac{1}{k}\ln\left(\frac{c}{c_0}\right)$$

Now, remember that

$$\log X^n = n \log X$$

from Section 5.4.2 on page 149, so when $n = -1$,

$$\log X^{-1} = -\log X$$

so

$$t = -\frac{1}{k} \ln\left(\frac{c}{c_0}\right)$$

$$= \frac{1}{k} \ln\left(\frac{c_0}{c}\right)$$

Substituting $t = t_{\frac{1}{2}}$ and $c = \frac{1}{2} c_0$ now gives

$$t_{\frac{1}{2}} = \frac{1}{k} \ln\left(\frac{c_0}{\frac{1}{2} c_0}\right)$$

$$= \frac{\ln 2}{k}$$

because the c_0 terms cancel, and $1 \div \frac{1}{2} = 2$.

Worked example 5.11

A determination of the pressure of gaseous N_2O_5 as a function of time gave the values

$\dfrac{t}{s}$	0	1200	4800	7200
$\dfrac{p}{mm\ Hg}$	350	190	34	10

If the decomposition of N_2O_5 is a first-order reaction, what is the value of the rate constant?

Chemical background

This is a frequently quoted example of a first-order reaction, the equation for the decomposition being

$$2N_2O_{5(g)} \rightarrow 4NO_{2(g)} + O_{2(g)}$$

Solid N_2O_5 consists of NO_2^+ and NO_3^- ions. It is colourless and stable below 0°C. At higher temperatures it decomposes slowly into N_2O_4 and O_2.

The total pressure can be measured by connecting the flask to a manometer, but it is then necessary to calculate the contribution from N_2O_5.

Pressure is a useful quantity to determine because it is proportional to concentration. If we consider the ideal gas equation

$$pV = nRT$$

and divide both sides by V, we obtain

$$p = \left(\frac{n}{V}\right) RT$$

The quantity n/V has the units of mole per unit volume, so is actually a measure of concentration c. We can therefore write

$$p = cRT$$

and because R is a constant at a constant temperature, p will be proportional to c.

Solution to worked example 5.11

Because this is a first-order reaction, we will use the integrated rate equation

$$\ln c - \ln c_0 = -kt$$

If the concentration c is proportional to pressure p, we can write

$$c = Kp$$

and

$$c_0 = Kp_0$$

where K (in this case equal to $1/RT$) is the constant of proportionality as explained in Chapter 3, Section 3.7.3, and p_0 is the initial pressure. Substituting into the integrated rate equation then gives

$$\ln(Kp) - \ln(Kp_0) = -kt$$

We saw in Section 5.4.2 on page 149 that

$$\ln(XY) = \ln X + \ln Y$$

and so

$$\ln(Kp) = \ln K + \ln p$$

and

$$\ln(Kp_0) = \ln K + \ln p_0$$

can be substituted into the left-hand side of the equation to give

$$(\ln K + \ln p) - (\ln K + \ln p_0) = -kt$$

The brackets can be removed, ensuring that the minus sign is applied to both terms in the second bracket, to give

$$\ln K + \ln p - \ln K - \ln p_0 = -kt$$

The terms $\ln K$ and $-\ln K$ cancel, to leave

$$\ln p - \ln p_0 = -kt$$

Adding $\ln p_0$ to both sides of the equation gives

$$\ln p = -kt + \ln p_0$$

Because p_0 is a constant, $\ln p_0$ is also a constant, and so we now have an equation that is of the form $y = mx + c$, as discussed in Chapter 3, Section 3.6.2, on page 78. Comparing corresponding quantities in the two forms of the equation shows that if $\ln p$ is plotted on the y-axis and t on the x-axis, we will obtain a straight-line graph with gradient $-k$.

The mechanics of graph plotting were outlined in Chapter 3, Section 3.6.3, on page 80. We first need to draw up a table containing values of $\ln p$ and t. We will actually calculate the values of $\ln(p/\text{mm Hg})$ because we can only take the logarithm of a quantity that has no units. This actually has the effect of introducing a further constant but does not affect the value of the gradient.

The first value in the table has

$$\frac{p}{\text{mm Hg}} = 350$$

so, if we use a calculator, we obtain

$$\ln\left(\frac{p}{\text{mm Hg}}\right) = \ln 350$$

$$= 5.86$$

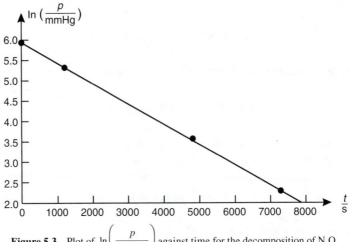

Figure 5.3   Plot of $\ln\left(\dfrac{p}{\text{mm Hg}}\right)$ against time for the decomposition of N_2O_5.

We use the [ln] button to evaluate the natural logarithms.

We can obtain other values in a similar fashion, and the data we need to plot can be tabulated as

$\dfrac{t}{\text{s}}$	0	1200	4800	7200
$\ln\left(\dfrac{p}{\text{mm Hg}}\right)$	5.86	5.25	3.53	2.30

The graph of $\ln\left(\dfrac{p}{\text{mm Hg}}\right)$ against t is shown in Figure 5.3, from which we can obtain the gradient.

$$\text{gradient} = \frac{\text{increase in } \ln\left(\dfrac{p}{\text{mm Hg}}\right)}{\text{increase in } t}$$

$$= \frac{2.00 - 5.85}{(7900 - 0)\text{s}}$$

$$= \frac{-3.85}{7900\,s}$$

$$= -4.9 \times 10^{-4}\,\text{s}^{-1}$$

We have already seen that the gradient is equal to $-k$, so

$$k = 4.9 \times 10^{-4}\ \text{s}^{-1}$$

5.5   Second-order reactions

If we consider the generalized expression for the rate v of a reaction

$$v = k[A]^x[B]^y$$

there are two ways in which it is possible to achieve an overall order of 2, i.e., to satisfy the equation

$$x + y = 2$$

Case 1: x = 2, y = 0

If $x = 2$ and $y = 0$ (or $x = 0$, and $y = 2$), the rate equation is

$$v = k[A]^2$$

This would arise in a reaction of the type

$$A + B \rightarrow products$$

in which the starting concentrations of A and B were equal.

Case 2: x = y = 1

If $x = y = 1$, the equation is

$$v = k[A][B]$$

The first of these two cases is rather simpler to deal with than is the second.

Worked example 5.12

The rate equation for a second-order reaction in which the initial concentrations of the reactants are equal can be expressed as

$$-\frac{dc}{dt} = kc^2$$

where dc/dt is the rate of change of concentration c of a reactant with time t, and k is the rate constant. Obtain an expression for the concentration c at a time t, assuming that the initial concentration value is c_0.

Chemical background

Typical reactions that obey second-order kinetics are

$$H_{2(g)} + I_{2(g)} \rightarrow 2HI_{(g)}$$

and

$$CH_3Br + OH^- \rightarrow CH_3OH + Br^-$$

The second of these is an example of an S_N2 (substitution nucleophilic bimolecular) reaction, which involves a transition state in which carbon is partially bonded to both −OH and −Br, as shown in Figure 5.4.

Figure 5.4   Transition state in the S_N2 reaction.

Solution to worked example 5.12

The initial stages involve rearranging the given equation to separate the terms, i.e., to have all terms in c on the left-hand side and all those in t on the right-hand side. Then we can integrate and assign limits as in worked examples 5.4 and 5.8. This results in us being required to solve the equation

$$\int_{c_o}^{c} \frac{dc}{c^2} = -k \int_{0}^{t} dt$$

The integration on the right-hand side is straightforward and identical to the previous problems.

$$\int_{0}^{t} dt = t$$

Integration of $1/c^2$ requires a little more thought but is not difficult once we realize that $1/c^2$ can be written as c^{-2}. Raising the power by one gives us c^{-1} (not c^{-3}, which would be a reduction of the power by one — this is a common error), so we need to divide c^{-1} by -1. Note that dividing by -1 is the same as multiplying by -1 because

$$x \div -1 = \frac{x}{-1} = -x \quad \text{and} \quad x \times -1 = -x$$

Consequently, we need to evaluate the expression

$$\left[-\frac{1}{c} \right]_{c_0}^{c} = -k \left[t \right]_{0}^{t}$$

If we substitute the upper limits and then subtract the value obtained with the lower limits, we get

$$-\frac{1}{c} - \left(-\frac{1}{c_0}\right) = -k(t-0)$$

which, if we multiply every term by −1, becomes

$$\frac{1}{c} - \frac{1}{c_0} = kt$$

Worked example 5.13

What are the units of the second-order rate constant?

Chemical background

An example of the second-order rate constant is the value of 2.42×10^{-2} $dm^3\ mol^{-1}\ s^{-1}$ for the reaction

$$H_2 + I_2 \rightarrow 2HI$$

Solution to worked example 5.13

As with other integrated rate equations, we need to rearrange it to make the rate constant k the subject of the equation. We can do this by dividing both sides of the equation by t, to give

$$k = \frac{1}{t}\left(\frac{1}{c} - \frac{1}{c_0}\right)$$

The units we would obtain from substituting into this equation would therefore be

$$\frac{1}{s}\left(\frac{1}{mol\,dm^{-3}}\right)$$

Note that

$$\frac{1}{s} = s^{-1} \quad \text{and} \quad \frac{1}{dm^{-3}} = dm^3$$

Then the units can be expressed as $dm^3\ mol^{-1}\ s^{-1}$. We write the units with the positive powers before those with the negative powers because, when expressed verbally, we replace negative powers by the word "per." Thus, $dm^3\ mol^{-1}\ s^{-1}$ would be spoken as decimeters cubed per mole per second.

An alternative method for synthesizing hydrogen iodide is by reducing iodine with hydrogen sulfide. Hydrogen compounds generally become less stable as a group of the periodic table is descended, so HF will be more stable than HI.

Worked example 5.14

If the concentration of a reactant in a second-order reaction, with equal concentrations of reactants, was monitored as a function of time, what graph would you plot to obtain the value of the rate constant?

Chemical background

The reactions of the carbonyl compounds acetaldehyde, propionaldehyde, and acetone with HCN to give cyanohydrins were found to be second order in a study reported in the *Journal of the American Chemical Society* in 1953 (**75**, 3106). Measurements were performed in acetate buffers under various conditions to ascertain the influence of ionic strength, pH, and buffer composition on the rate constants.

Solution to worked example 5.14

In order to obtain a value of the rate constant k, we need to plot the data in such a way that we obtain a straight line. This can be done by rearranging the integrated rate equation above into the form of the equation of a straight-line graph, $y = mx + c$, as outlined in Chapter 3, Section 3.6.2, on page 78.

Starting with the equation

$$\frac{1}{c} - \frac{1}{c_0} = kt$$

we can add the constant term $1/c_0$ to both sides to give

$$\frac{1}{c} = kt + \frac{1}{c_0}$$

This is now in the form we want, with $1/c$ corresponding to y and t corresponding to x. If a graph of $1/c$ against t is plotted, we will obtain a straight line with gradient k and intercept $1/c$ as shown in Figure 5.5.

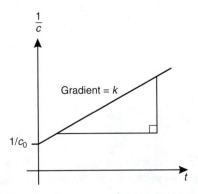

Figure 5.5   Graph of integrated rate equation for a second-order reaction.

Worked example 5.15

During the hydrolysis of methyl acetate, aliquots of the reaction mixture were withdrawn at intervals and titrated with sodium hydroxide to give the following results:

$\dfrac{t}{\text{min}}$	5.64	20.69	45.74	75.76
$\dfrac{V}{\text{cm}^3}$	26.52	27.98	29.88	31.99

where V is the titre of sodium hydroxide. Determine the order of this reaction.

Chemical background

The data given refer to the acid hydrolysis of methyl acetate, which can be represented by the equation

$$CH_3COOCH_3 + H_2O = CH_3COOH + CH_3OH$$

The carbonyl oxygen becomes protonated and so the carbonyl carbon atom is more susceptible to nucleophilic attack by the water molecule. There are likely to be several tetrahedral intermediates in this reaction.

Solution to worked example 5.15

We have seen earlier that there are three distinct plots that will give a straight line in reactions we are likely to meet. To summarize, we need to plot

- c against t for a zero-order reaction
- $\ln c$ against t for a first-order reaction
- $1/c$ against t for a second-order reaction

To determine the order of a reaction such as this, we need to perform the three plots and identify which one gives the straight line. Hopefully, this will be easy to spot.

In this example, there are further complications. We are not given the concentration of methyl acetate but the titre of sodium hydroxide. However, because the rate of increase of methanoic acid is numerically equivalent to the rate of decrease of methyl acetate, these quantities are proportional, and we can write

$$V = Kc$$

where K is the constant of proportionality, and so

$$V_0 = Kc_0$$

Methyl acetate is an ester that can be formed by the reaction of acetic anhydride with methanol. It melts at −98°C and boils at 57.5°C.

If we now substitute into the integrated rate equation for a zero-order reaction

$$c = -kt + c_0$$

we have

$$\frac{V}{K} = -kt + \frac{V_0}{K}$$

We can multiply both sides by the proportionality constant K to give

$$V = -kKt + V_0$$

so, in fact, a plot of V against t will be a straight line having gradient kK and intercept V_0.

In a similar way, it is possible to show that straight-line plots will be obtained for a reaction of appropriate order if a quantity that is proportional to the concentration is used instead of the concentration itself. In this problem, it is therefore possible to use the titre values directly rather than converting them to concentrations.

A further complication is that we do not have the values of V_0, the titre value at zero time. However, as the time values are all relative, we can subtract the first value of 5.64 from each, which is equivalent to assuming that the reaction began 5.64 min later. If we do this, the data we need to use become

$\dfrac{t}{\text{min}}$	0.00	15.05	40.10	70.12
$\dfrac{V}{\text{cm}^3}$	26.52	27.98	29.88	31.99

To consider the first- and second-order plots, we need the values of

$$\ln\left(\frac{V}{\text{cm}^3}\right) \quad \text{and} \quad \frac{1}{V}$$

If we consider the first point, we have

$$V = 26.52 \text{ cm}^3$$

$$\frac{V}{\text{cm}^3} = 26.52$$

$$\ln\left(\frac{V}{\text{cm}^3}\right) = 3.278$$

$$\frac{1}{V} = 0.03771 \text{cm}^{-3}$$

The full table with these quantities is as follows:

t/min	0.00	15.05	40.10	70.12
V/cm^3	26.52	27.98	29.88	31.99
$\ln(V/\text{cm}^3)$	3.278	3.331	3.397	3.465
V^{-1}/cm^{-3}	0.03771	0.03574	0.03346	0.03126

Use the **ln** button to evaluate the natural logarithms, and the x^{-1} button to evaluate the reciprocals.

The graphs of V against t, $\ln V$ against t, and $1/V$ against t are shown in Figure 5.6. Since only the plot of $\ln V$ against t gives a straight line, the reaction must be first order.

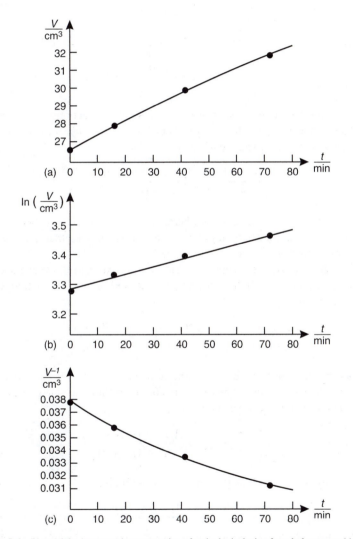

Figure 5.6 Plots of the integrated rate equations for the hydrolysis of methyl acetate: (a) zero-order, (b) first-order, and (c) second-order.

To analyze second-order kinetics in the situation when the starting concentrations of the two reactants are not equal, we need some additional mathematical techniques.

5.5.1   Partial fractions

Partial fractions are useful for expressing a fractional product as the sum of two fractions, which in some situations may be easier to deal with. For example, if we have an expression such as

$$\frac{1}{(x+1)(x+2)}$$

it may be preferable to express it as the sum of two fractions such as

$$\frac{A}{x+1} + \frac{B}{x+2}$$

where A and B are constants. Determining the partial fractions, in this case, is a matter of finding values for the constants A and B. Because these expressions are identical, we can write the *identity*

$$\frac{1}{(x+1)(x+2)} \equiv \frac{A}{(x+1)} + \frac{B}{(x+2)}$$

Notice the use of the $\equiv$ symbol here. It is used instead of $=$ because this expression is true for all values of x. An equation, on the other hand, is only true for certain values of x.

If we now consider the right-hand side of this expression, we can multiply the top and bottom of the first term by $(x+2)$ and the top and bottom of the second term by $(x+1)$:

$$\frac{1}{(x+1)(x+2)} \equiv \frac{A(x+2)}{(x+1)(x+2)} + \frac{B(x+1)}{(x+2)(x+1)}$$

because multiplying both the top and bottom of a fraction by the same quantity does not change its value. The two terms on the right-hand side now have the same denominators, so we can now write

$$\frac{1}{(x+1)(x+2)} \equiv \frac{A(x+2) + B(x+1)}{(x+1)(x+2)}$$

Multiplying both expressions by $(x+1)(x+2)$ gives

$$1 \equiv A(x+2) + B(x+1)$$

Now, because this is an identity, it will be true for any value of x. If we choose to set $x = -1$, the bracket containing $x+1$ will become zero, and we have

$$A(-1 + 2) \equiv 1$$

$$A \equiv 1$$

Similarly, if we set $x = -2$, the bracket containing $x + 2$ becomes zero, and we have

$$B(-2 + 1) \equiv 1$$

$$-B \equiv 1$$

$$B \equiv -1$$

Substituting into our general expression for the partial fractions then gives

$$\frac{1}{(x+1)(x+2)} \equiv \frac{1}{x+1} + \frac{-1}{x+2}$$

$$\equiv \frac{1}{x+1} - \frac{1}{x+2}$$

There are more complicated cases of partial fractions, but this knowledge is sufficient for our present purposes.

5.5.2   *Differentiation of logarithmic functions and integration of fractions*

We have seen previously in Section 5.4.1 on page 149 that

$$\int \frac{dx}{x} = \ln x + C$$

where C is the constant of integration. If integration of $1/x$ gives $\ln x$, and differentiation is the reverse of integration (Section 5.3.1 on page 141), differentiation of $\ln x$ will give $1/x$, expressed mathematically as

$$\frac{d}{dx}(\ln x) = \frac{1}{x}$$

It is also true that

$$\frac{d}{dx}(\ln(a - x)) = \frac{1}{x - a}$$

so that, for example

$$\frac{d}{dx}(\ln(3 - x)) = \frac{1}{x - 3}$$

Because integration is the reverse of differentiation, we also have

$$\int \frac{dx}{x-a} = \ln(a-x) + C$$

where C is the constant of integration.

There is a complication in that we can only take the logarithm of a positive number, and the correct expression for dealing with general integrations of this type is

$$\int \frac{dx}{ex+f} = \frac{1}{e}\ln|ex+f| + C$$

The vertical bars denote that we take the modulus of the quantity enclosed between them. This is the value taking no account of its sign, as we saw previously in Chapter 4, Section 4.4, on page 120. In the examples we meet in chemistry, the modulus symbols are frequently omitted, so we would write

$$\int \frac{dx}{ex+f} = \frac{1}{e}\ln(ex+f) + C$$

Worked example 5.16

The rate equation for a second-order reaction in which the initial concentrations of the reactants are different can be expressed as

$$-\frac{da}{dt} = kab$$

where da/dt is the rate of change of concentration a of one reactant with time t, b is the concentration of the second reactant, and k is the rate constant. Obtain an expression for the concentrations a and b at a time t assuming that the initial concentration values are a_0 and b_0.

Chemical background

The analysis of general second-order reactions of this type is quite complicated. It is also possible to have reactions of higher order, but the analysis of these will not be considered here as they are rarely encountered. One example of a reaction exhibiting third-order kinetics is that between nitric oxide and chlorine

$$2NO + Cl_2 \rightarrow 2NOCl$$

which obeys the rate equation

$$-\frac{d[NO]}{dt} = k[NO]^2[Cl_2]$$

Bacterial action is responsible for 80% of NO in the atmosphere, whereas the rest is produced by combustion of various types. Both NO and NO_2 are toxic and take part in photochemical reactions in the atmosphere — they are known as NO_x gases.

Solution to worked example 5.16

As previously, we begin by separating the variables, i.e., rearranging the expression given so that all the concentration terms are on the left, with the rate constant and the time on the right. This gives us the expression

$$\int \frac{da}{ab} = -k \int dt$$

where the rate constant k is taken outside the integral on the right-hand side. We immediately have a problem because we need to integrate

$$\frac{1}{ab}$$

with respect to a, and this cannot be done as the expression stands because it also contains the variable b. To overcome this difficulty, remember that the generalized equation for this reaction is

$$A + B \rightarrow \text{products}$$

The initial concentrations of A and B are given as a_0 and b_0, respectively. Suppose that, after time t, the concentration of A has reduced by x; if A and B react in an equimolar ratio, then the concentration of B will also have been reduced by x, and we can write for the concentrations of A and B

$$a = a_0 - x$$

and

$$b = b_0 - x$$

Because a_0 and b_0 are constants, we have now expressed the concentrations of A and B in terms of a single common variable x. If we differentiate the first equation, because a_0 is a constant, then

$$\frac{da}{dx} = -1$$

and so we can replace da by $-dx$ in the integration.

We are now able to substitute for a, b, and da to obtain

$$\int \frac{dx}{(a_0 - x)(b_0 - x)} = k \int dt$$

noticing that the minus signs cancel on the right-hand side. This expression now needs to be integrated between appropriate limits. Clearly, when $t = 0$, none of A or B has reacted, so $x = 0$. We want to obtain an expression giving x after time t so we use these values as our upper limits:

$$\int_0^x \frac{dx}{(a_0 - x)(b_0 - x)} = k \int_0^t dt$$

Integrating the right-hand side is straightforward:

$$\int_0^t dt = \left[t \right]_0^t = t - 0 = t$$

However, we now have the problem of integrating the expression on the left, which is not so easy. We saw earlier how an expression of this type could be written in terms of two partial fractions. As before, we start by setting up an identity in terms of two unknown constants A and B:

$$\frac{1}{(a_0 - x)(b_0 - x)} \equiv \frac{A}{a_0 - x} + \frac{B}{b_0 - x}$$

We multiply the first term, top and bottom, by $(b_0 - x)$ and the second term, top and bottom, by $(a_0 - x)$; neither of these operations changes the value of the expression, but it does produce fractions with the same denominators:

$$\frac{1}{(a_0 - x)(b_0 - x)} \equiv \frac{A(b_0 - x)}{(a_0 - x)(b_0 - x)} + \frac{B(a_0 - x)}{(b_0 - x)(a_0 - x)}$$

Because the quantity on the bottom is the same for every term in the expression, we multiply all terms by $(a_0 - x)(b_0 - x)$ to give

$$1 \equiv A(b_0 - x) + B(a_0 - x)$$

This expression is true for all values of x, so we set $x = b_0$ to make the first bracket zero and to leave

$$1 \equiv B(a_0 - b_0)$$

This can be rearranged to give

$$B \equiv \frac{1}{a_0 - b_0}$$

Similarly, setting $x = a_0$ makes the second bracket zero, and we have

$$1 \equiv A(b_0 - a_0)$$

which can be rearranged to give

$$A \equiv \frac{1}{b_0 - a_0}$$

It is worth noticing that $A = -B$, as this will allow some simplification later.
Substituting in our original identity for the partial fractions gives

$$\frac{1}{(a_0 - x)(b_0 - x)} \equiv \frac{1}{(b_0 - a_0)(a_0 - x)} + \frac{1}{(a_0 - b_0)(b_0 - x)}$$

If we multiply the top and bottom of the second term on the right by -1,
this becomes

$$\frac{1}{(a_0 - x)(b_0 - x)} \equiv \frac{1}{(b_0 - a_0)(a_0 - x)} - \frac{1}{(b_0 - a_0)(b_0 - x)}$$

We can then introduce brackets on the right-hand side and take the common factor of $(b_0 - a_0)$ outside the new brackets to give

$$\frac{1}{(a_0 - x)(b_0 - x)} = \frac{1}{(b_0 - a_0)}\left(\frac{1}{(a_0 - x)} - \frac{1}{(b_0 - x)} \right)$$

Bearing in mind that $b_0 - a_0$, and consequently $\dfrac{1}{b_0 - a_0}$ is a constant,
we can now write our equation to be integrated as

$$\frac{1}{b_0 - a_0} \int_0^x \left(\frac{1}{(a_0 - x)} - \frac{1}{(b_0 - x)} \right) dx = k \int_0^t dt$$

or

$$\frac{1}{b_0 - a_0} \int_0^x \frac{dx}{a_0 - x} - \frac{1}{b_0 - a_0} \int_0^x \frac{dx}{b_0 - x} = k \int_0^t dt$$

We saw earlier that

$$\int \frac{dx}{ex + f} = \frac{1}{e}\ln(ex + f) + C$$

Comparing this with the first integral

$$\int \frac{dx}{a_0 - x}$$

shows that $e = -1$ and $f = a_0$. Consequently,

$$\int \frac{dx}{a_0 - x} = \frac{1}{-1} \ln(-x + a_0)$$

$$= -\ln(a_0 - x) + C$$

This allows us to write the following expression for the definite integral:

$$\int_0^x \frac{dx}{a_0 - x} = \left[-\ln(a_0 - x) \right]_0^x$$

Substituting for the upper limit and then subtracting the expression with the lower limit gives

$$\int_0^x \frac{dx}{a_0 - x} = \left[-\ln(a_0 - x) \right] - \left[-\ln(a_0 - 0) \right]$$

$$= -\ln(a_0 - x) + \ln a_0$$

because the adjacent negative signs combine to give a positive one. We saw in Section 5.4.2 on page 149 that

$$\ln D - \ln E = \ln\left(\frac{D}{E} \right)$$

so that

$$\int_0^x \frac{dx}{a_0 - x} = \ln\left(\frac{a_0}{a_0 - x} \right)$$

Similarly, we obtain

$$\int_0^x \frac{dx}{b_0 - x} = \ln\left(\frac{b_0}{b_0 - x} \right)$$

and finally we have

$$\int_0^t dt = t$$

as we have already seen several times for the other rate equations. Putting the results of these three integrations into the overall expression leads to

$$\frac{1}{b_0 - a_0}\left(\ln\left(\frac{a_0}{a_0 - x}\right) - \ln\left(\frac{b_0}{b_0 - x}\right) \right) = kt$$

One again we have the difference of two logarithmic terms, so we can write this as

$$\frac{1}{b_0 - a_0}\left(\ln\left(\frac{a_0}{a_0 - x}\right)\left(\frac{b_0}{b_0 - x}\right)^{-1} \right) = kt$$

which can be written as

$$\frac{1}{b_0 - a_0}\left(\ln\frac{a_0(b_0 - x)}{b_0(a_0 - x)} \right) = kt$$

because

$$\left(\frac{b_0}{b_0 - x}\right)^{-1} = \frac{b_0 - x}{b_0}$$

Our integrations above have been performed with respect to the variable x because this could be related to the concentrations of both A and B. However, it is now possible to eliminate x from the expression because we set

$$a = a_0 - x$$

and

$$b = b_0 - x$$

where a and b represent concentrations after time t. This gives the expression as

$$\frac{1}{b_0 - a_0} \ln\left(\frac{a_0 b}{a b_0}\right) = kt$$

Because

$$\left(\frac{a_0 b}{a b_0}\right) = \left(\frac{a b_0}{a_0 b}\right)^{-1}$$

this could also be written as

$$\frac{1}{b_0 - a_0} \ln\left(\frac{a b_0}{a_0 b}\right)^{-1} = kt$$

From Section 5.4.2 on page 149 we know that

$$\ln x^{-1} = -\ln x$$

and so

$$\frac{-1}{b_0 - a_0} \ln\left(\frac{a b_0}{a_0 b}\right) = kt$$

Multiplying this expression top and bottom by –1 gives the final expression

$$\frac{1}{a_0 - b_0} \ln\left(\frac{a b_0}{a_0 b}\right) = kt$$

The derivation of this expression requires several mathematical techniques and not a few "tricks of the trade" to get intermediate expressions into the required form; fortunately, few mathematical processes we have met or will meet are of such length or complexity!

5.6   The Arrhenius equation

In fact, such a law can also be used to describe other processes, one of these being the chirping of crickets. This may seem surprising at first, until we realize that the chirping is itself controlled by a chemical reaction.

So far we have considered the rate constant k to be constant under specified conditions. This is in fact true as long as the temperature remains constant but, as you might expect, the rate of a reaction increases with the temperature. The relationship between rate constant and temperature is described by an exponential function.

5.6.1   The exponential function

The exponential function can be defined in terms of our usual notation as

$$f(x) = e^x$$

where e has the value 2.7183 to four decimal places. This value is not exact, but we rarely need to worry about the numerical value of e. When complicated expressions are involved, you will also see the exponential function written as

$$f(x) = \exp(x)$$

This has precisely the same meaning as the previous equation and is merely a useful device for improving the clarity of some mathematical expressions.

Scientific calculators allow easy calculation of exponential functions.

It is worth noting that it is only possible to take the exponential of a quantity that has no units; in other words, a pure number. The absence of units in an expression should always be verified before attempting to calculate an exponential term in a calculation.

A typical sequence of buttons would be followed by in order to evaluate the exponential function.

5.6.2   Inverse functions

The inverse of a function $f(x)$ is generally known as $\mathrm{arc}f(x)$ and reverses the effect of the original function. For example, if a function $f(x)$ is defined by

$$f(x) = x + 10$$

this means "take a value of x and add 10 to it." To reverse this we must therefore subtract 10, and we write

$$\mathrm{arc}f(x) = x - 10$$

Notice that

$$\mathrm{arc}f(f(x)) = f(x) - 10$$

$$= x + 10 - 10$$

$$= x$$

which is a general result for a function and its inverse.

The natural logarithmic and exponential functions are actually the inverses of each other, so that

$$\mathrm{arc}\ \ln(x) = e^x, \text{ and arc } e^x = \ln x$$

It also follows that

$$e^{\ln x} = x, \text{ and } \ln e^x = x$$

Similarly,

$$10^{\log x} = x \text{ and } \log 10^x = x$$

In other words, the inverse of $\log x$ is 10^x.

Worked example 5.17

The value of the rate constant k for the decomposition of nitrogen dioxide was found to vary with absolute temperature T according to the Arrhenius equation

$$k = A \exp\left(\frac{-E_a}{RT}\right)$$

NO$_2$ can be formed in the laboratory by the catalytic oxidation of ammonia gas by oxygen from the air, followed by the exothermic oxidation of nitric oxide to nitrogen dioxide.

where A is a constant called the preexponential factor, E_a is the activation energy, and R is the gas constant that can be taken as 8.31 J K^{-1} mol^{-1}. Use the data below to determine the value of the activation energy E_a:

$\dfrac{T}{K}$	593	604	628	652	657
$\dfrac{k}{dm^3\,mol^{-1}\,s^{-1}}$	0.523	0.7511	1.70	4.01	5.01

Chemical background

For some reactions involving free atoms or radicals, there is a very small activation energy, and a more accurate treatment of temperature dependence must be used. This frequently involves using an equation of the form

$$k = AT^n \exp\left(-\frac{E}{RT}\right)$$

where the value of n depends on the nature of the reaction and the theory being used in the analysis. The activation energy E in this equation is related to that in the Arrhenius equation, E_a, by the equation

$$E = E_a - nRT$$

Solution to worked example 5.17

A straightforward plot of k against T is shown in Figure 5.7(a) and is obviously curved. As usual, we would like to obtain a linear plot, and this can be done by taking the natural logarithm of each side of the Arrhenius equation. This gives us

$$\ln k = \ln\left(A \exp\left(-\frac{E_a}{RT}\right)\right)$$

The right-hand side of this equation now has the logarithm of a product, and because

$$\ln(XY) = \ln X + \ln Y$$

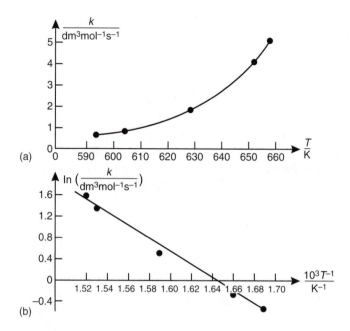

Figure 5.7 Plot of (a) rate constant k against temperature T, and (b) ln k against $1/T$ for the decomposition of nitrogen dioxide.

as we saw in section 5.4.2 on page 149, we can rewrite the equation as

$$\ln k = \ln A + \ln\left(\exp\left(-\frac{E_a}{RT}\right)\right)$$

We have also seen that $\ln e^x = x$, and so our modified Arrhenius equation becomes

$$\ln k = \ln A - \frac{E_a}{RT}$$

Reversing the order of the terms on the right and grouping the constants E_a and R in brackets gives us

$$\ln k = -\left(\frac{E_a}{R}\right)\left(\frac{1}{T}\right) + \ln A$$

which is equivalent to the general equation of a straight line $y = mx + c$, with ln k being equivalent to y, $-E_a/R$ being equivalent to the gradient m, $1/T$ being equivalent to x, and ln A being equivalent to the intercept c. We therefore need to plot ln k against $1/T$ to obtain a straight line with gradient $-E_a/R$.

Because we can only take the logarithm of a pure number without units, we will actually calculate values of

$$\ln\left(\frac{k}{\text{dm}^3\,\text{mol}^{-1}\,\text{s}^{-1}}\right)$$

Strictly speaking, this is the quantity that should be on the left-hand side of the equation above. However, it is tedious to keep writing this, so we tend to use the equation as shown and remember that we need to deal with the units in this way. For the first point,

$$k = 0.523 \text{ dm}^3\,\text{mol}^{-1}\,\text{s}^{-1}$$

so

$$\frac{k}{\text{dm}^3\,\text{mol}^{-1}\,\text{s}^{-1}} = 0.523$$

and using a calculator gives

$$\ln\left(\frac{k}{\text{dm}^3\,\text{mol}^{-1}\,\text{s}^{-1}}\right) = -0.648$$

Press **ln** to obtain the value of the natural logarithm.

The temperature values are given in units of K, so directly give the values of T. The first point has $T = 593$ K, and the calculator gives

$$T^{-1} = \frac{1}{T} = 1.69 \times 10^{-3}\,K^{-1}$$

The x^{-1} key gives the value of the reciprocal.

It is worth looking at how this expression can be rearranged to assist in the tabulation of data. It is neater to incorporate the power of ten into the table heading, rather than including it for every value. The remaining values of T^{-1} are calculated in the same way and can be tabulated as follows.

$\ln\left(\dfrac{k}{\text{dm}^3\,\text{mol}^{-1}\,\text{s}^{-1}}\right)$	−0.648	−0.286	0.531	1.39	1.61
$\dfrac{10^3\,T^{-1}}{K^{-1}}$	1.69	1.66	1.59	1.53	1.52

The plot of $\ln\left(\dfrac{k}{\text{dm}^3\,\text{mol}^{-1}\,\text{s}^{-1}}\right)$ against T^{-1} is shown in Figure 5.7(b).

From this, we can calculate

$$\text{gradient} = \frac{\text{increase in } \ln k}{\text{increase in } T^{-1}}$$

$$= \frac{-0.65 - 1.85}{(1.70 - 1.50) \times 10^{-3} \text{K}^{-1}}$$

$$= \frac{-2.50}{0.20 \times 10^{-3} \text{K}^{-1}}$$

$$= -1.25 \times 10^4 \text{K}$$

We saw earlier that the gradient is equal to $-\dfrac{E_a}{R}$ so that

$$-\frac{E_a}{R} = -1.25 \times 10^4 \text{K}$$

The negative signs on either side of the equation cancel, and multiplying both sides by the gas constant R gives

$$E_a = 1.25 \times 10^4 \text{ K} \times R$$

Substituting the value of R gives

$$E_a = 1.25 \times 10^4 \text{ K} \times 8.31 \text{ J K}^{-1} \text{ mol}^{-1}$$

$$= 1.04 \times 10^5 \text{ J mol}^{-1}$$

As we might expect, the activation energy has units of J mol^{-1}. Because 1 kJ = 10^3 J, and $10^5 = 10^3 \times 10^2$, we also have

$$E_a = 1.04 \times 10^2 \text{ kJ mol}^{-1}$$

$$= 104 \text{ kJ mol}$$

Worked example 5.18

The activation energy for the reaction

$$2 \text{ H}_2\text{O}_2 = 2\text{H}_2\text{O} + \text{O}_2$$

is 48.9 kJ mol^{-1} in the presence of a colloidal platinum catalyst. What is the effect on the rate constant of raising the temperature from 20°C to 30°C?

Hydrogen peroxide can be
produced industrially by the
partial oxidation of 2-ethyl-
anthraquinol or 2-propanol
by air. It can be extracted
into water.

Chemical background

This is another example of a disproportionation reaction. If traces of certain ions are present, these catalyze the reaction, which then occurs rapidly. One such ion is Fe^{3+}, which may alternate with Fe^{2+} while the reaction is taking place.

Solution to worked example 5.18

If we assume that the rate constant has a value of k_1 when the absolute temperature is T_1, and a value of k_2 at temperature T_2, we can write the logarithmic form of the Arrhenius equation for the two cases as:

$$\ln k_1 = \ln A - \frac{E_a}{RT_1} \quad \text{and} \quad \ln k_2 = \ln A - \frac{E_a}{RT_2}$$

Notice that the preexponential factor is constant for a given reaction, so the value of A is the same in both equations.

Subtracting the first of these equations from the second gives

$$\ln k_2 - \ln k_1 = \ln A - \frac{E_a}{RT_2} - \left(\ln A - \frac{E_a}{RT_1} \right)$$

$$= \ln A - \ln A + \frac{E_a}{RT_1} - \frac{E_a}{RT_2}$$

$$= \frac{E_a}{RT_1} - \frac{E_a}{RT_2}$$

$$= \frac{E_a}{R} \left(\frac{1}{T_1} - \frac{1}{T_2} \right)$$

Because

$$\ln X - \ln Y = \ln \left(\frac{X}{Y} \right)$$

as we saw in Section 5.4.2 on page 149, this then becomes

$$\ln \left(\frac{k_2}{k_1} \right) = \frac{E_a}{R} \left(\frac{1}{T_1} - \frac{1}{T_2} \right)$$

and from the left-hand side of the equation we can obtain the ratio k_2/k_1, which indicates the change in the rate constant relative to its initial value k_1.

We need to convert our temperature values in °C to absolute temperature values in K. This gives

$$T_1 = (20 + 273) = 293 \text{ K}$$

and

$$T_2 = (30 + 273) = 303 \text{ K}$$

so that we can now calculate the quantity enclosed in brackets in the equation on the previous page, which is

$$\left(\frac{1}{T_1} - \frac{1}{T_2} \right) = \frac{1}{293\text{K}} - \frac{1}{303\text{K}}$$

Evaluating these reciprocals using a calculator gives

$$\left(\frac{1}{T_1} - \frac{1}{T_2} \right) = 3.41 \times 10^{-3} \text{K}^{-1} - 3.30 \times 10^{-3} \text{K}^{-1}$$

$$= (3.41 - 3.30) \times 10^{-3} \text{K}^{-1}$$

$$= 0.11 \times 10^{-3} K^{-1}$$

$$= 1.1 \times 10^{-4} K^{-1}$$

Substituting these values into our equation gives

$$\ln\left(\frac{k_2}{k_1} \right) = \frac{E_a}{R} \left(\frac{1}{T_1} - \frac{1}{T_2} \right)$$

$$= \frac{48.9 \, \text{kJ mol}^{-1}}{8.314 \, \text{J K}^{-1} \text{mol}^{-1}} \times 1.1 \times 10^{-4} \, \text{K}^{-1}$$

$$= \frac{48.9 \times 10^3 \, \text{J mol}^{-1}}{8.314 \, \text{J K}^{-1} \text{mol}^{-1}} \times 1.1 \times 10^{-4} \, \text{K}^{-1}$$

$$= 0.647$$

Notice that the units cancel to give a value for $\ln\left(\dfrac{k_2}{k_1} \right)$ without units, as required. Because

$$\ln\left(\frac{k_2}{k_1} \right) = 0.647$$

To do this press

followed by .

we need to take the exponential function of both sides of this equation to remove the logarithmic term and give

$$\frac{k_2}{k_1} = e^{0.647}$$

and with the use of a calculator we obtain

$$\frac{k_2}{k_1} = 1.91$$

This reaction has an activation energy that is fairly typical for reactions that proceed at a reasonable rate. A useful rule of thumb is that a temperature increase of 10°C will produce a doubling of the reaction rate.

We see that for this reaction an increase in temperature of 10°C produces almost a doubling of the rate constant.

5.7   The steady-state approximation

This provides us with a useful means of deriving rate equations for reactions that proceed via the formation of one or more intermediate species. The concentration of such a species is assumed to increase rapidly at the start of a reaction, remain relatively constant during most of the reaction, and then decrease rapidly at the end of the reaction. The period when the concentration is constant corresponds to a zero gradient on a graph of concentration against time, as shown in Figure 5.8. As we saw in Chapter 4, Section 4.6.2, on page 127, this corresponds to a zero value of the derivative, i.e.,

$$\frac{dc}{dt} = 0$$

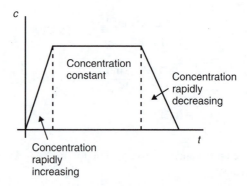

Figure 5.8   Schematic graph of the variation in concentration c of a reactive intermediate with time t.

5.7.1   Simultaneous equations

If we have a single equation, then the most we can ever hope to achieve is to solve it to give the value of a single variable. If we have more than one variable, then the only way we can determine their values is if we have more information. In fact, we need at least as many equations as there are unknowns; it is also the case that such equations must be independent of each other.

For example, if we have the equations

$$x - y = 2 \qquad\qquad (1)$$

and

$$2x + 3y = 9 \qquad\qquad (2)$$

which are true at the same time, then it follows that the same value of x and the same value of y must satisfy each one at that time.

It is usual to label simultaneous equations as we have done here with (1) and (2). Generally, we can solve a pair of simultaneous equations by adding or subtracting them, or by adding or subtracting a multiple of one or both of them, so as to eliminate one of the variables.

In this example we will choose to eliminate y. To do this, we first need to multiply equation (1) throughout by 3; we will call this equation (3):

$$3x - 3y = 6 \qquad\qquad (3)$$

We now add equation (2) and equation (3). To do this we need to treat each side separately, so we obtain

$$2x + 3y + 3x - 3y = 9 + 6$$

This can be simplified to give

$$5x = 15$$

and dividing both sides by 5 gives

$$x = 3$$

We can now substitute in either of our original equations. If we choose equation (1), we have

$$3 - y = 2$$

and adding y to each side of this gives

$$3 = 2 + y$$

from which, subtracting 2 from either side, gives

$$y = 1$$

As a check we can substitute these values of x and y into equation (2). On the left-hand side of this we have

$$2x + 3y = (2 \times 3) + (3 \times 1) = 6 + 3 = 9$$

which is the required value on the right-hand side.

Worked example 5.19

In the reaction

$$C_2H_6 \rightarrow C_2H_4 + H_2$$

which gives ethene from ethane, the two intermediates are the radicals CH_3, C_2H_5, and H. Applying the steady-state approximation to these gives the simultaneous equations

$$\frac{d[CH_3]}{dt} = 2k_1[C_2H_6] - k_2[CH_3][C_2H_6] = 0 \tag{1}$$

$$\frac{d[C_2H_5]}{dt} = k_2[CH_3][C_2H_6] - k_3[C_2H_5] + k_4[H][C_2H_5] - 2k_5[C_2H_5]^2 = 0 \tag{2}$$

$$\frac{d[H]}{dt} = k_3[C_2H_5] - k_4[H][C_2H_6] = 0 \tag{3}$$

Solve these equations to give an expression for $[C_2H_5]$ in terms of $[C_2H_6]$.

Chemical background

The preceding equations are based on the fact that this reaction proceeds by means of a chain reaction. It begins with an initiation step involving the formation of radicals. This can happen in two ways, represented by the equations

$$C_2H_6 \rightarrow 2\,CH_3$$

and

$$CH_3 + C_2H_6 \rightarrow CH_4 + C_2H_5$$

for which the rate constants are denoted as k_1 and k_2, respectively. The propagation steps

$$C_2H_5 \rightarrow C_2H_4 + H$$

$$H + C_2H_6 \rightarrow H_2 + C_2H_5$$

can continue in a cyclical fashion to give the products of the reaction. The rate constants of these steps are k_3 and k_4, respectively. The final step is that of termination:

$$2\,C_2H_5 \rightarrow C_4H_{10}$$

for which the rate constant is k_5.

Solution to worked example 5.19

Because we require an answer in terms of $[C_2H_5]$ and $[C_2H_6]$, we need to eliminate the terms $[CH_3]$ and $[H]$ from the scheme of the preceding equations.

We can eliminate $[CH_3]$ by adding equations (1) and (2), and we can eliminate $[H]$ by adding equations (2) and (3). Consequently, we can eliminate both $[CH_3]$ and $[H]$ by adding (1), (2), and (3). This gives

$$2\,k_1\,[C_2H_6] - 2\,k_5\,[C_2H_5]^2 = 0$$

Adding $2\,k_5\,[C_2H_5]^2$ to both sides gives

$$2\,k_1\,[C_2H_6] = 2\,k_5\,[C_2H_5]^2$$

Dividing both sides by $2\,k_5$ gives

$$[C_2H_5]^2 = \frac{k_1[C_2H_6]}{k_5}$$

Taking the square root of each side then gives

$$[C_2H_5] = \sqrt{\frac{k_1[C_2H_6]}{k_5}}$$

Exercises

1. The variable y has the value 17.2 when x is 3.4, and the value 29.7 when x is 10.1. What is the average rate of change of y with respect to x?

2. Determine the following indefinite integrals:

   (a) $\int x^5 dx$

   (b) $\int 7x^6 dx$

   (c) $\int x^{-4} dx$

3. Determine these indefinite integrals:

(a) $\int (6x^2 + 9x + 8)dx$

(b) $\int (3x^3 + 4x^2)dx$

(c) $\int (3x^4 - 4x^2 + 6)dx$

4. Calculate these definite integrals:

(a) $\int_{-1}^{1} (x^3 + x^2 + x)dx$

(b) $\int_{0}^{3} (4x^2 + 2x + 1)dx$

(c) $\int_{-2}^{4} (4x^7 - 6x^5 - 9x^3)dx$

5. Determine these integrals:

(a) $\int \left(x + \dfrac{1}{x} \right) dx$

(b) $\int_{1}^{2} \left(2x^2 + \dfrac{3}{x} \right) dx$

(c) $\int_{3}^{6} \left(8x^3 - \dfrac{6}{x} \right)$

6. Determine:

(a) $\int \dfrac{dx}{x^3}$

(b) $\int_{1}^{5} \dfrac{4}{x^6} dx$

(c) $\int_{-3}^{5} \dfrac{dx}{x^8}$

7. Determine:

(a) $\int \left(3x^2 + \dfrac{5}{x} + \dfrac{2}{x^3} \right) dx$

(b) $\int_{2}^{10} \left(4x^5 - \dfrac{3}{x+2} + \dfrac{4}{x^5} \right) dx$

(c) $\int \left(9x^7 - \dfrac{4}{3x-1} - \dfrac{3}{x^6} \right) dx$

8. Simplify each of these expressions into a single logarithm:

(a) $\log 3 + \log 4$

(b) $\ln 1 + \ln 2 + \ln 3$

(c) $2 \ln 4 - 3 \ln 2$

9. Express

$$\frac{x}{(x+2)(x+3)}$$

in terms of its partial fractions.

10. Differentiate these expressions:

(a) $2x^2 + 3 \ln x$

(b) $3x + \ln(x-2)$

(c) $4x^3 - \ln(2x+1)$

11. Integrate these expressions:

(a) $\int \left(x + \dfrac{1}{x-2} \right) dx$

(b) $\int_{0}^{1} \left(2x^2 + \dfrac{1}{x-5} \right) dx$

(c) $\int_{3}^{7} \left(\dfrac{2}{3x+1} - \dfrac{7}{4x-3} \right) dx$

12. Evaluate the integral

$$\int_5^6 \frac{dx}{(x-4)(x+5)}$$

13. Calculate these quantities:

(a) $2e^{-3}$

(b) $4e^2$

(b) $3e^0$

14. Simplify the following expressions:

(a) $(e^2)^3$

(b) e^4e^5

(c) $\dfrac{x^2e^6}{e^3}$

15. Solve each of the following equations for x:

(a) $10^x = 42$

(b) $e^x = 0.75$

(c) $e^{2x-1} = 3.2$

16. Solve each of the following equations for x:

(a) $\ln x = 3.6$

(b) $\ln 4x = 7.2$

(c) $\log(2x + 3) = 6.1$

17. Determine the inverses of these functions:

(a) $f(x) = 3x^2 + 5$

(b) $f(x) = 2\sqrt{x-8}$

(c) $f(x) = 2 + \dfrac{7}{8x^2}$

18. Solve the simultaneous equations

$$4x + 2y = 5$$

$$3x + y = 9$$

19. Solve the simultaneous equations

$$3x + 4y = 6$$

$$x - 2y = -8$$

20. Solve the simultaneous equations

$$3x + 2y + 5z = 2$$

$$5x + 3y - 2z = 4$$

$$2x - 5y - 3z = 14$$

Problems

1. Azomethane at an initial pressure of 0.080 mm Hg was allowed to decompose at 350°C. After an hour, its partial pressure was 0.015 mm Hg. What is the average rate of the reaction?

2. Trichloroamine reacts with liquid or concentrated aqueous HCl according to the equation

$$NCl_3 + 4HCl \rightarrow NH_4Cl + 3Cl_2$$

Write down a series of relationships between the rates of change of the concentrations of each of the four chemical species.

3. For a reaction involving reactant A having a rate defined by the equation

$$-\frac{d[A]}{dt} = k[A]^a$$

with k being the rate constant and a some number that may or may not be an integer, obtain the integrated rate equation for the concentration [A] after time t in terms of $[A]_o$, the value when $t = 0$. Are there any values of a for which this equation is not valid?

4. It is reasonable to suppose that the rate v of the reaction

$$2\,NO_{(g)} + Cl_{2(g)} \rightarrow 2\,NOCl_{(g)}$$

is given by an equation of the form

$$v = k\,[NO]^m[Cl_2]^n$$

where k is the rate constant. Obtain the logarithmic form of this equation and explain how this could be used to obtain the values of m and n.

5. The concentration of glucose in aqueous hydrochloric acid was monitored and found to give the following results:

t	c
min	$10^2\,\mathrm{mol\,dm^{-3}}$
120	5.94
240	5.65
360	5.43
480	5.15

Determine the order of the reaction and the rate constant.

6. At 20°C the rate constant for a reaction is $2.7 \times 10^3\,\mathrm{s^{-1}}$. What is the value of the rate constant at 50°C if the activation energy is 22.3 kJ mol^{-1}?

7. Use the Arrhenius equation to calculate the activation energy for a reaction whose rate constant increases by a factor of 3 when the temperature is increased from 20°C to 30°C.

8. The Eyring equation gives the rate constant k for a reaction in terms of the Boltzmann constant k_B and Planck constant h as

$$k = \frac{k_B T}{h} e^{-\frac{\Delta G^\circ}{RT}}$$

where T is the absolute temperature, R the gas constant, and ΔG° the Gibbs energy of activation. Calculate ΔG° for the reaction

$$H_2 + I_2 \rightarrow 2\,HI$$

for which the value of the rate constant is 0.0234 dm^3 mol^{-1} s^{-1} at 400°C.

9. For the reaction between hydrogen and bromine, application of the steady-state approximation gives the equations

$$\frac{d[H]}{dt} = k_2[Br][H_2] - k_3[H][Br_2] - k_4[H][HBr] = 0$$

$$\frac{d[Br]}{dt} = 2k_1[Br_2] - k_2[Br][H_2] + k_3[H][Br_2]$$

$$+ k_4[H][HBr] - 2k_5[Br]^2 = 0$$

Solve these equations to give an expression for [Br] in terms of [Br$_2$].

10. For the decomposition of N_2O_5 the following series of equations is obtained:

$$\frac{d[NO_3]}{dt} = k_1[N_2O_5] - k_{-1}[NO_2][NO_3]$$

$$- k_2[NO_2][NO_3] - k_3[NO][NO_3] = 0$$

$$\frac{d[NO]}{dt} = k_2[NO_2][NO_3] - k_3[NO][NO_3] = 0$$

Solve these equations to give an expression for [NO] in terms of $[NO_2]$.

Structural chemistry 6

Although the structural properties of solids, liquids, and gases are all important, in this chapter we will be concerned mainly with the structures of solids, and particularly those solids that are crystalline. One of the reasons for this is that far more information is available on the structures of crystalline solids than any other form of matter, it being obtained by the technique of x-ray crystallography.

The concept of symmetry is very important when we are studying structural properties, and we will also look at the use of simple symmetry operators. A more detailed treatment of symmetry involves a reasonable knowledge of group theory, which will not be covered here because an elementary treatment relevant to chemists is usually included in the appropriate section of physical chemistry textbooks.

6.1 Packing fractions of atoms in metals

We can think of a metal crystal as comprising a series of spherical atoms packed together. There are various ways in which the atoms may be packed together, and these give rise to different types of structure. One way in which these differ is in the volume of space actually occupied by atoms; this is known as the packing fraction f_v, and is defined by

$$f_v = \frac{\text{volume occupied by atoms}}{\text{volume of unit cell}}$$

where the unit cell is the unit that defines the structure, and which is repeated by symmetry.

Worked example 6.1

In a cubic close-packed structure the volume of atoms is determined as $4 \times \frac{4}{3}\pi r^3$ and the volume of the unit cell as $(r\sqrt{8})^3$, where r is the radius of the atom. In body-centred cubic packing, the atoms have volume $2 \times \frac{4}{3}\pi r^3$, and the volume of the unit cell is $\left(\frac{4r}{\sqrt{3}}\right)^3$. What is the ratio of the packing fractions for the two cases?

(a)

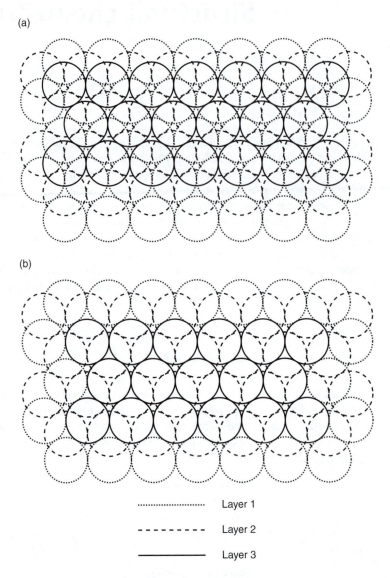

(b)

.... Layer 1

- - - - - - - Layer 2

———————— Layer 3

Figure 6.1   Arrangement of atoms in (a) cubic close-packed and (b) hexagonal close-packed crystals.

Chemical background

The most efficient way of arranging spherical atoms is closest packing. It is possible to have both cubic close packing and hexagonal close packing. Examples of cubic close-packed structures are calcium, nickel, and copper, whereas magnesium, cobalt, and zinc are hexagonally close packed. On the other hand, some alkali metals have a body-centred cubic structure, as do iron and chromium (Figure 6.1).

Solution to worked example 6.1

For the cubic close-packed structure we have

$$f_v = \frac{\text{volume occupied by atoms}}{\text{volume of unit cell}}$$

$$= \frac{4 \times \frac{4}{3}\pi r^3}{(r\sqrt{8})^3}$$

The denominator of this expression can be rewritten as

$$(r\sqrt{8})^3 = r^3 8^{\frac{3}{2}}$$

because, using the rules of indices introduced in Chapter 1, Section 1.5, on page 13

$$\sqrt{8} = 8^{\frac{1}{2}}$$

and

$$(8^{\frac{1}{2}})^3 = 8^{\frac{3}{2}}$$

Performing the multiplication in the numerator then leads to

$$f_v = \frac{\frac{16}{3}\pi r^3}{r^3 8^{\frac{3}{2}}}$$

The term r^3 now cancels, we can substitute for the value of π, and the use of a calculator finally gives

The ⌃ calculator button can be used to raise numbers to fractional as well as integral powers.

$$f_v = \frac{\frac{16}{3}\pi}{8^{\frac{3}{2}}} = \frac{16.757}{22.627} = 0.741$$

There are slightly more elegant ways to obtain this solution, but as we end up with a decimal fraction, we are vindicated in using a calculator at the earliest stage possible.

For the body-centred cubic structure

$$f_v = \frac{\text{volume occupied by atoms}}{\text{volume of unit cell}}$$

$$= \frac{2 \times \frac{4}{3}\pi r^3}{\left(\frac{4r}{\sqrt{3}}\right)^3}$$

From Chapter 1, Section 1.4.4, on page 12, we recall that dividing by a fraction is the same as multiplying by its reciprocal. Consequently

$$f_v = 2 \times \frac{4}{3}\pi r^3 \times \left(\frac{\sqrt{3}}{4r}\right)^3 = \frac{2 \times 4\pi r^3 \times 3\sqrt{3}}{3 \times 4^3 r^3}$$

$$= \frac{2\pi\sqrt{3}}{4^2} = 0.680$$

because

$$\left(\sqrt{3}\right)^3 = \sqrt{3} \times \sqrt{3} \times \sqrt{3} = 3\sqrt{3}$$

and several terms cancel top and bottom.

The ratio

$$\frac{f_v(\text{close packing})}{f_v(\text{body-centered cubic packing})} = \frac{0.741}{0.680} = 1.090$$

Note that because there is more empty space in body-centred cubic structures, these tend to be more malleable and less brittle.

6.2   Arrangement of atoms in crystals

Crystals are characterized by having atoms in fixed average positions, even though the constant movement of the atoms means that, at any instant, a particular atom may be slightly displaced. Consequently, it is possible to define interatomic distances with constant values within a crystal.

In the previous section we met the concept of a unit cell. The unit cell is defined by three edges of lengths a, b, and c, and three angles α, β, and

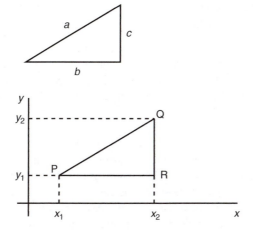

Figure 6.2   The relationship described by Pythagoras' theorem.

γ between these edges. The position of an atom within the unit cell can be defined in terms of its position relative to the three edges; these are known as fractional coordinates. It is also possible to define the position of an atom with respect to three axes at right angles, and it is usual to denote these as x, y, and z, with these quantities being actual rather than fractional lengths.

Here we will use the latter system. It is possible to convert from one to the other, but this is a specialized operation that is beyond the scope of this book.

6.2.1   Pythagoras' theorem

If we have a right-angled triangle such as the one shown in Figure 6.2, we can identify the two sides b and c that define the right angle together with the third longer side a, which is known as the hypotenuse. The lengths of these sides are related by the equation

$$a^2 = b^2 + c^2$$

or, by taking the square root of both sides

$$a = \sqrt{b^2 + c^2}$$

We can now place this triangle within a two-dimensional coordinate system and label the three corners of the triangle as P, Q, and R. The point P is assumed to have coordinates (x_1, y_1), and Q is assumed to have coordinates (x_2, y_2). It follows from this that R must have coordinates (x_2, y_1) as shown in Figure 6.2, that length PR is $x_2 - x_1$, and RQ is $y_2 - y_1$. The distance between P and Q is then given by Pythagoras' theorem as

$$\sqrt{(x_2 - x_1)^2 + (y_2 - y_1)^2}$$

Thus, in a planar molecule, if we can assign x and y coordinates to a pair of atoms, we can calculate the distance between them.

Worked example 6.2

A carbon atom in a planar alkene molecule has coordinates (0.250, 0.694), whereas a hydrogen atom to which it is bonded has coordinates (−0.295, 0.250). What is the carbon–hydrogen bond length? The coordinates are given in Å.

Chemical background

A carbon atom in a double carbon–carbon bond occupies a planar environment due to sp^2 hybridization. This ensures that electrons in all the bonds are as far apart as possible and results in bond angles of 120°. In practice it would not be possible to obtain atomic positions in this molecule from x-ray crystallography; ethane freezes at −183°C and is not a crystalline solid.

Solution to worked example 6.2

Because we have the coordinates of both atoms, we can apply Pythagoras' theorem. If we assign

$$x_1 = 0.250 \qquad y_1 = 0.694$$
$$x_2 = -0.295 \qquad y_2 = 0.250$$

We use to square the quantities, and once added, the button to take the square root.

we can substitute directly into the equation above to give the bond length r as

$$r = \sqrt{(x_2 - x_1)^2 + (y_2 - y_1)^2}$$
$$= \sqrt{(-0.295 - 0.250)^2 + (0.250 - 0.694)^2}$$
$$= \sqrt{(-0.545)^2 + (-0.444)^2}$$
$$= \sqrt{0.297 + 0.197}$$
$$= \sqrt{0.494}$$
$$= 0.702$$

Note that the answer will be in Å. Strictly speaking, the units should be included throughout the calculation, but here they are excluded for clarity. Notice also that squaring a negative number gives a positive number.

6.2.2 *Pythagoras' theorem in three dimensions*

Most molecules are not planar, and in such cases the position of an atom needs to be expressed in terms of three coordinates, usually x, y, and z. A direct extension of Pythagoras' theorem then gives the distance d between two atoms having coordinates (x_1, y_1, z_1) and (x_2, y_2, z_2), respectively, as

$$d = \sqrt{(x_2 - x_1)^2 + (y_2 - y_1)^2 + (z_2 - z_1)^2}$$

Worked example 6.3

The coordinates of two atoms in a metal complex are shown below. Determine the distance between the barium and nitrogen atoms.

	x	y	z
Ba	5.9000	–3.5985	6.6116
N	6.4452	–4.3097	9.2980

Chemical background

X-ray crystallography with single crystals gives the positions of atoms but no direct information on which atoms are bonded. This has to be inferred from the distance between atoms; bond lengths between specified types of atoms are relatively constant.

Solution to worked example 6.3

Substituting in the equation above gives

We again use the

 and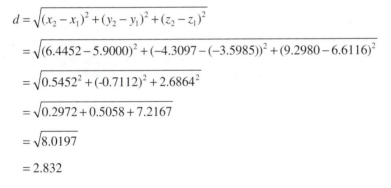

buttons to facilitate this calculation.

$$d = \sqrt{(x_2 - x_1)^2 + (y_2 - y_1)^2 + (z_2 - z_1)^2}$$

$$= \sqrt{(6.4452 - 5.9000)^2 + (-4.3097 - (-3.5985))^2 + (9.2980 - 6.6116)^2}$$

$$= \sqrt{0.5452^2 + (-0.7112)^2 + 2.6864^2}$$

$$= \sqrt{0.2972 + 0.5058 + 7.2167}$$

$$= \sqrt{8.0197}$$

$$= 2.832$$

because

$$-4.3097 - (-3.5985) = -4.3097 + 3.5985$$

and the square of a negative number is positive. Once again, the units have been omitted for clarity but are actually Å.

6.3   Bragg's law

Bragg's law is mathematically quite simple, and yet its study introduces many of the techniques that will be of use to us later on. It governs the behaviour of x-rays when they are diffracted from a crystal.

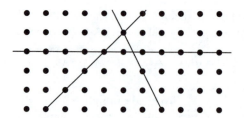

Figure 6.3   Planes in a crystal lattice.

The great advance brought about by the Bragg equation was that it considered radiation of a fixed wavelength. Previously, it had only been possible to use "white" x-rays, which were a mixture of various wavelengths and resulted in a far more complicated diffraction pattern.

Because crystals consist of units (atoms, ions, or molecules) that are arranged in an orderly fashion, they may be represented by a series of lattice points, each of which has the same environment. Various planes may be drawn through these points, as shown in Figure 6.3. Because the spacing of atoms in a crystal is similar to the wavelength of incident x-rays, diffraction rather than reflection occurs with such radiation from planes of atoms. The diffraction angle θ is therefore dependent upon the separation d of particular planes as well as on the wavelength λ of the incident x-rays.

6.3.1   Trigonometry

The trigonometric functions are evaluated using the , , and tan buttons. Prior to this the appropriate degrees or radian mode must be selected using ⬭.

The trigonometric functions are a set of functions that take as their input an angle. The simplest three are defined in terms of a right-angled triangle containing an angle θ, as shown in Figure 6.4. The lengths of the sides of the triangle are labeled as: opposite the angle of interest, adjacent to the angle of interest, and the hypotenuse (which is opposite the right angle). The sine, cosine, and tangent functions are then defined by the equations

$$\sin\theta = \frac{\text{opposite}}{\text{adjacent}}$$

$$\cos\theta = \frac{\text{adjacent}}{\text{hypotenuse}}$$

$$\tan\theta = \frac{\text{opposite}}{\text{adjacent}}$$

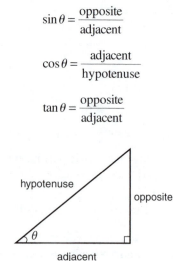

Figure 6.4   Definition of the trigonometric functions.

which show that each function is defined as the ratio of two lengths and therefore has no units.

Although angles are normally expressed in units of degrees, it is sometimes more convenient to express them in terms of the quantity radians, which has the symbol rad. These are related by the equation

$$2\pi\text{rad} = 360°$$

so that

$$\pi \text{ rad} = 180°$$

In practice, it is not usually necessary to convert from one to the other because scientific calculators have both "degrees" and "radians" mode. Make sure you are working in the correct one!

The graphs of the sine, cosine, and tangent functions are shown in Figure 6.5. It is worth remembering that

$$\sin 0° = \cos 90° = 0$$

$$\sin 90° = \cos 0° = 1$$

$$\tan 0° = \frac{\sin 0°}{\cos 0°} = \frac{0}{1} = 0$$

Worked example 6.4

The Bragg equation is expressed as

$$n\lambda = 2d\sin \theta$$

where n is called the order of reflection (having integer values of 1, 2, 3, etc.), and the other terms have been defined in the preceding text. Calculate the lattice spacing d when copper K_α radiation of wavelength 0.154 nm is incident on a cubic crystal and produces a first-order ($n = 1$) reflection with a scattering angle of 11°.

Chemical background

X-rays can be generated by firing high-energy electrons at a target, which is most frequently copper metal. Copper electrons are displaced from the innermost K shell and replaced by those from shells further out such as L and M, which release their excess energy as x-rays, as shown in Figure 6.6. When replacement of an electron occurs from the next level up L, the radiation is known as K_α, although it is also possible (but less likely) to obtain K_β when the replacement electron is from the next highest level M.

Other targets used for the generation of x-rays include molybdenum, cobalt, iron, and chromium.

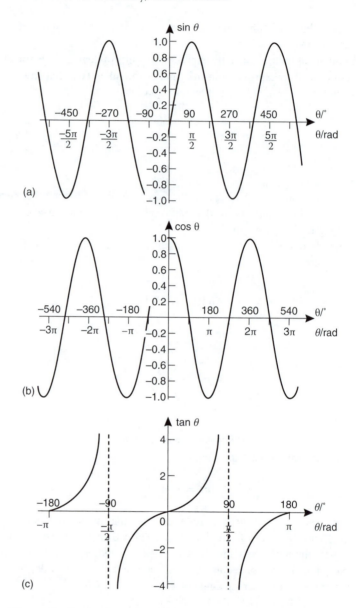

Figure 6.5   Graphs of trigonometric functions: (a) sin θ; (b) cos θ; and (c) tan θ.

Figure 6.6   Production of K_α x-rays.

Solution to worked example 6.4

To calculate the lattice spacing d, we need to rearrange the equation to make d the subject. We can do this by dividing both sides by $2 \sin \theta$:

$$d = \frac{n\lambda}{2 \sin \theta}$$

We have the quantities required to make a straightforward substitution into this equation, giving us

$$d = \frac{1 \times 0.154 \, \text{nm}}{2 \times \sin 11^\circ}$$

$$= \frac{0.154 \, \text{nm}}{2 \times 0.1908}$$

$$= 0.404 \, \text{nm}$$

We use the [sin] calculator button to evaluate this sine function.

Notice that in working through this problem we did not need to convert the units of the wavelength from nm to m, and the units of the spacing of the planes follow directly from substituting into the equation.

6.3.2   Inverses of trigonometric functions

The idea of inverse functions was introduced in Chapter 5, Section 5.6.2, on page 173. An inverse function effectively undoes the effect of the original function.

Each of the trigonometric functions we have met has its own inverse. These are

$$\arcsin x \text{ written as } \sin^{-1} x$$

$$\arccos x \text{ written as } \cos^{-1} x$$

$$\arctan x \text{ written as } \tan^{-1} x$$

The inverse trigonometric functions are evaluated using the combinations

[SHIFT] [sin],

[SHIFT] [cos], and

[SHIFT] [tan],

respectively.

The use of $^{-1}$ as a superscript in these expressions does *not* have any relationship to its use elsewhere to represent a reciprocal and is purely the nomenclature used to denote the inverse function.

To see how the inverse trigonometric functions are used in practice, consider the function defined as

$$f(x) = \sin(3x + 4)$$

When calculating the inverse of this function, we clearly need to generate x from $\sin x$, and this can only be done by using the inverse trigonometric function. We must therefore have the term $\sin^{-1} x$ in the expression for the inverse function $\text{arc}f(x)$. Having done that, we work through the operations in reverse order. In this case, that involves subtracting 4 and then dividing by 3 so that we obtain

$$arcf(x) = \frac{(\sin^{-1} x) - 4}{3}$$

Worked example 6.5

Potassium nitrate can be prepared by fractional crystallization from a solution of sodium nitrate and potassium chloride. It is used in gunpowder.

A certain set of lattice planes in potassium nitrate crystals has a spacing of 543 pm. Calculate the first-order ($n = 1$) diffraction angle when copper K_{α} radiation of wavelength 154 pm is incident on these planes.

Chemical background

The crystal structure of potassium nitrate consists of an orthorhombic lattice in which the angles are each 90°. The dimensions of the sides of the unit cell are 5.431 Å, 9.164 Å, and 5.414 Å, respectively.

Solution to worked example 6.5

It is probably easiest to determine the value of $\sin \theta$ and then to generate θ by using the inverse function. The first stage is to rearrange the Bragg equation to make $\sin \theta$ the subject. This can be done by dividing both sides by $2d$:

$$\sin \theta = \frac{n\lambda}{2d}$$

which, on substitution of the values given, leads to

$$\sin \theta = \frac{1 \times 154 \, pm}{2 \times 543 \, pm}$$

$$= 0.1418$$

We now need to apply the inverse sine function to both sides of this equation.

$$\sin^{-1}(\sin \theta) = \sin^{-1}(0.1418)$$

The calculator button sequence for this is

Notice that $\sin^{-1}(\sin \theta) = \theta$. Using a calculator for $\sin^{-1}(0.1418)$ we find

$$\theta = 8.15°$$

6.4   The unit cell

The monoclinic form of sulfur crystallizes in the space group known as P2$_1$/c.

The lattice points referred to in Section 6.3 on page 197 can be used to define the unit cell of a crystal. This is a three-dimensional unit from which it is

Figure 6.7   Example of a crystal lattice in two dimensions.

possible to generate the whole lattice (Figure 6.7). Unit cells may be primitive, having lattice points only at their corners, or nonprimitive, in which case additional lattice points are present.

6.4.1   Unit vectors

Vectors are numbers that are associated with a particular direction. Because of this they are very useful for dealing with problems in three dimensions. Any vector can be defined in terms of its components in three directions, normally chosen to be mutually at right angles and denoted as *x, y,* and *z. Unit vectors,* having a size of one unit, can be defined in each of the directions and are denoted as **i, j**, and **k**, respectively. Note the use of bold typescript to denote vector quantities; when written by hand they can simply be underlined. These unit vectors can be combined by simple addition, so that a vector **a** might be defined as

$$\mathbf{a} = 2\mathbf{i} + 3\mathbf{j} + \mathbf{k}$$

This vector is shown relative to the three axes *x, y,* and z in Figure 6.8.

The preceding definition of a vector is of one that can be thought of as starting at the origin and whose finish is found by moving 2 units in the *x*-direction,

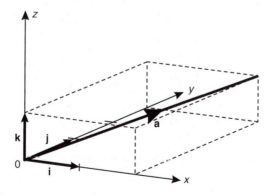

Figure 6.8   The vector $2\mathbf{i} + 3\mathbf{j} + \mathbf{k}$.

3 units in the y-direction, and 1 unit in the z-direction. The actual vector is then represented diagrammatically by a straight line starting at the origin and finishing at this second point, as in Figure 6.8.

6.4.2   Addition and subtraction of vectors

Vectors can be added and subtracted using the standard rules of arithmetic applied to each direction separately. For example, if vectors **b** and **c** are defined as

$$\mathbf{b} = 2\mathbf{i} + 3\mathbf{j}$$

$$\mathbf{c} = 2\mathbf{j} + \mathbf{k}$$

then

$$\mathbf{b} + \mathbf{c} = 2\mathbf{i} + 3\mathbf{j} + 2\mathbf{j} + \mathbf{k}$$

$$= 2\mathbf{i} + (3\mathbf{j} + 2\mathbf{j}) + \mathbf{k}$$

$$= 2\mathbf{i} + 5\mathbf{j} + \mathbf{k}$$

Worked example 6.6

Figure 6.9 shows lattice points situated at the corners of a primitive orthogonal lattice having dimensions a, b, and c; the origin O and axes x, y, and z are shown. In terms of the unit vectors **i**, **j**, and **k**, determine the position vector of each point A–G.

Chemical background

Such a crystal lattice, containing points only at the corners of each unit cell, is known as a primitive lattice. Because it has the form of a cuboid if a, b, and c are different, it is known as orthorhombic. An example of a structure having a primitive orthorhombic unit cell is that of the mineral Forsterite, $Mg_2(SiO_4)$.

Figure 6.9   The primitive orthogonal lattice.

Solution to worked example 6.6

We need to consider each lattice point in turn and define the path that takes us from the origin to that point.

(a) To move from O to A, we need to go a units in the x-direction. Because the unit vector in this direction is **i**, the vector we require is $a\mathbf{i}$.

(b) Point B lies on the y-axis at a distance of b units from O. The unit vector along this axis is **j**, so the required vector is $b\mathbf{j}$.

(c) Point C lies on the z-axis, along which the unit vector is **k**. We need to move c units, so this point is defined by vector $c\mathbf{k}$.

(d) To get from O to D we need to move a units in the x-direction and b units in the y-direction. The vector is consequently $a\mathbf{i} + b\mathbf{j}$.

(e) Point E is reached by moving a units in the x-direction and c units in the z-direction. No movement in the y-direction is required. The defining vector is therefore $a\mathbf{i} + c\mathbf{k}$.

(f) Point F requires movement of b units in the y-direction and c units in the z-direction, so the required vector is $b\mathbf{j} + c\mathbf{k}$.

(g) Point G is the only one that requires a movement in each of the three directions; a units in x, b units in y, and c units in z. The vector defining this point is consequently $a\mathbf{i} + b\mathbf{j} + c\mathbf{k}$.

6.4.3   Multiplication of vectors

There are two ways in which a pair of vectors can be multiplied together. One results in a pure number, also known as a scalar quantity, whereas the other gives another vector. The first method is known as calculating the *scalar product*, or dot product, because this is the symbol used to denote the process. For example, the scalar product of two vectors **a** and **b** would be denoted as $a \cdot b$ and is defined as

$$a \cdot b = |\mathbf{a}||\mathbf{b}|\cos\theta$$

where $|\mathbf{a}|$ and $|\mathbf{b}|$ are known as the *moduli* of **a** and **b**, respectively, and θ is the angle between them.

The modulus or magnitude of a vector is its length (a pure number or scalar quantity) and can be calculated by taking the square root of the sums of the squares of the contributions of each unit vector to the total. In two dimensions, this is Pythagoras' theorem that we met in Section 6.2.2, on page 196. In three dimensions, for example, if

$$\mathbf{v} = 2\mathbf{i} + \mathbf{j} - 3\mathbf{k}$$

then

$$|v| = \sqrt{(2^2 + 1^2 + (-3)^2)}$$

$$= \sqrt{(4+1+9)}$$

$$= \sqrt{14}$$

The calculator functions

and can be used to facilitate this calculation.

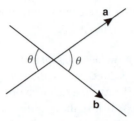

Figure 6.10   Definition of the angle θ between two vectors.

There are two possible angles that can be taken as θ because cos θ = cos($180° - \theta$), and your calculator will offer the acute angle θ. The correct one is shown in Figure 6.10, which shows that both vectors must be pointing away from, or both must be pointing towards, the point of intersection. The cosine of this angle, known as cos θ, is easily calculated using a scientific calculator, but it is important to ensure that it has been set to the appropriate mode to accept angles in degrees as input, and that you choose the correct angle θ, or $180° - \theta$. Notice also that $\mathbf{a} \cdot \mathbf{b} = \mathbf{b} \cdot \mathbf{a}$; it does not matter which order you take dot products.

If we consider the scalar products of the various combinations of unit vectors, we obtain a very useful set of relationships. For example,

$$\mathbf{i} \cdot \mathbf{i} = |\mathbf{i}||\mathbf{i}| \cos \theta$$

Now, $|\mathbf{i}|$ = 1, and the angle between $\mathbf{i}$ and $\mathbf{i}$ is zero, so $\theta = 0°$. Because cos $0°$ = 1, we then have

$$\mathbf{i} \cdot \mathbf{i} = 1$$

The same set of arguments leads to the relationships

$$\mathbf{j} \cdot \mathbf{j} = 1 \quad \text{and} \quad \mathbf{k} \cdot \mathbf{k} = 1$$

On the other hand, if we consider a scalar product such as $\mathbf{i} \cdot \mathbf{j}$, we then have

$$\mathbf{i} \cdot \mathbf{j} = |\mathbf{i}||\mathbf{j}| \cos \theta$$

where again $|\mathbf{i}|$ = 1 and $|\mathbf{j}|$ = 1, but this time the angle between $\mathbf{i}$ and $\mathbf{j}$ is a right angle, so $\theta = 90°$. Because cos $90° = 0$, we now have $\mathbf{i} \cdot \mathbf{j} = 0$. Notice then that

$$\mathbf{i} \cdot \mathbf{j} = \mathbf{j} \cdot \mathbf{i} = 0$$

Similarly

$$\mathbf{j} \cdot \mathbf{k} = \mathbf{k} \cdot \mathbf{j} = 0$$

and

$$\mathbf{i} \cdot \mathbf{k} = \mathbf{k} \cdot \mathbf{i} = 0$$

Now consider what happens if we need to calculate the scalar product of two vectors that are defined in terms of the unit vectors $\mathbf{i}$, $\mathbf{j}$, and $\mathbf{k}$, such as

$$\mathbf{a} = 3\mathbf{i} + 2\mathbf{j} + \mathbf{k}, \text{ and } \mathbf{b} = 2\mathbf{i} - \mathbf{j} + \mathbf{k}$$

We can expand this product as we would with any pair of brackets:

$$\mathbf{a} \cdot \mathbf{b} = (3\mathbf{i} + 2\mathbf{j} + \mathbf{k}) \cdot (2\mathbf{i} - \mathbf{j} + \mathbf{k})$$
$$= 3\mathbf{i} \cdot (2\mathbf{i} - \mathbf{j} + \mathbf{k}) + 2\mathbf{j} \cdot (2\mathbf{i} - \mathbf{j} + \mathbf{k}) + \mathbf{k} \cdot (2\mathbf{i} - \mathbf{j} + \mathbf{k})$$
$$= 6\mathbf{i} \cdot \mathbf{i} - 3\mathbf{i} \cdot \mathbf{j} + 3\mathbf{i} \cdot \mathbf{k} + 4\mathbf{j} \cdot \mathbf{i} - 2\mathbf{j} \cdot \mathbf{j} + 2\mathbf{j} \cdot \mathbf{k} + 2\mathbf{k} \cdot \mathbf{i} - \mathbf{k} \cdot \mathbf{j} + \mathbf{k} \cdot \mathbf{k}$$

Because

$$\mathbf{i} \cdot \mathbf{j} = \mathbf{j} \cdot \mathbf{i} = \mathbf{j} \cdot \mathbf{k} = \mathbf{k} \cdot \mathbf{j} = \mathbf{i} \cdot \mathbf{k} = \mathbf{k} \cdot \mathbf{i} = 0$$

and

$$\mathbf{i} \cdot \mathbf{i} = \mathbf{j} \cdot \mathbf{j} = \mathbf{k} \cdot \mathbf{k} = 1$$

this leaves us with

$$\mathbf{a} \cdot \mathbf{b} = 6 - 2 + 1 = 5$$

As well as calculating the scalar product of a pair of vectors, there are occasions when we need to multiply vectors and retain some of the information about direction. To do this, we can calculate the *vector product* of a pair of vectors. This is denoted as $\mathbf{a} \times \mathbf{b}$, and is defined as

$$\mathbf{a} \times \mathbf{b} = |\mathbf{a}|\,|\mathbf{b}|\sin\theta\,\hat{\mathbf{n}}$$

where the symbols have the same meaning as in the case of the scalar product. Notice this time, however, that the vector product is defined in terms of a vector $\hat{\mathbf{n}}$.

This gives the direction of the resulting vector and is itself of magnitude 1. $\hat{\mathbf{n}}$ is perpendicular to both the vectors $\mathbf{a}$ and $\mathbf{b}$, and its direction (up or down) can be found by using the corkscrew rule. If we imagine a right-handed corkscrew that rotates from vector $\mathbf{a}$ to vector $\mathbf{b}$, then the direction of travel of the corkscrew gives the direction of the vector $\hat{\mathbf{n}}$.

Note that

$$\mathbf{a} \times \mathbf{b} \neq \mathbf{b} \times \mathbf{a}$$

because the corkscrew rule results in the vector product pointing in opposite directions.

However

$$\mathbf{a} \times \mathbf{b} = -\mathbf{b} \times \mathbf{a}$$

We can apply this rule to calculate the vector products between pairs of the unit vectors $\mathbf{i}$, $\mathbf{j}$, and $\mathbf{k}$. Because the angle between pairs of unit vectors is $0°$, $\sin\theta = 0$, and so the vector products between a pair of identical vectors must be zero. This gives

$$\mathbf{i} \times \mathbf{i} = \mathbf{j} \times \mathbf{j} = \mathbf{k} \times \mathbf{k} = 0$$

where the zero is technically a vector quantity. We also know that sin 90° is 1, so, for a pair of perpendicular vectors, we do not need to worry about this term, and we again obtain a unit vector. If **i**, **j**, and **k** form a right-handed set so that the corkscrew rule can be applied, we obtain

$$\mathbf{i} \times \mathbf{j} = \mathbf{k}$$

$$\mathbf{k} \times \mathbf{i} = \mathbf{j}$$

$$\mathbf{j} \times \mathbf{k} = \mathbf{i}$$

Because $\mathbf{a} \times \mathbf{b} = -\mathbf{b} \times \mathbf{a}$, we also have

$$\mathbf{j} \times \mathbf{i} = -\mathbf{k}$$

$$\mathbf{i} \times \mathbf{k} = -\mathbf{j}$$

$$\mathbf{k} \times \mathbf{j} = -\mathbf{i}$$

This information allows us to calculate the vector product of any pair of vectors that are expressed in terms of the unit vectors **i**, **j**, and **k**. Using the example just presented, where

$$\mathbf{a} = 3\mathbf{i} + 2\mathbf{i} + \mathbf{k}$$

$$\mathbf{b} = 2\mathbf{i} - \mathbf{j} + \mathbf{k}$$

the expansion of the brackets gives

$$\mathbf{a} \times \mathbf{b} = (3\mathbf{i} + 2\mathbf{j} + \mathbf{k}) \times (2\mathbf{i} - \mathbf{j} + \mathbf{k})$$

$$= 3\mathbf{i} \times (2\mathbf{i} - \mathbf{j} + \mathbf{k}) + 2\mathbf{j} \times (2\mathbf{i} - \mathbf{j} + \mathbf{k}) + \mathbf{k} \times (2\mathbf{i} - \mathbf{j} + \mathbf{k})$$

$$= 3\mathbf{i} \times 2\mathbf{i} - 3\mathbf{i} \times \mathbf{j} + 3\mathbf{i} \times \mathbf{k} + 2\mathbf{j} \times 2\mathbf{i} - 2\mathbf{j} \times \mathbf{j}$$

$$+ 2\mathbf{j} \times \mathbf{k} + 2\mathbf{k} \times \mathbf{i} - \mathbf{k} \times \mathbf{j} + \mathbf{k} \times \mathbf{k}$$

Now

$$\mathbf{i} \times \mathbf{i} = \mathbf{j} \times \mathbf{j} = \mathbf{k} \times \mathbf{k} = 0$$

This leaves

$$\mathbf{a} \times \mathbf{b} = -3\mathbf{i} \times \mathbf{j} + 3\mathbf{i} \times \mathbf{k} + 4\mathbf{j} \times \mathbf{i} + 2\mathbf{j} \times \mathbf{k} + 2\mathbf{k} \times \mathbf{i} - \mathbf{k} \times \mathbf{j}$$

We can now substitute for each of these individual vector products to obtain

$$\mathbf{a} \times \mathbf{b} = -3\mathbf{k} + 3(-\mathbf{j}) + 4(-\mathbf{k}) + 2\mathbf{i} + 2\mathbf{j} - (-\mathbf{i})$$

$$= -3\mathbf{k} - 3\mathbf{j} - 4\mathbf{k} + 2\mathbf{i} + 2\mathbf{j} + \mathbf{i}$$

$$= 3\mathbf{i} - \mathbf{j} - 7\mathbf{k}$$

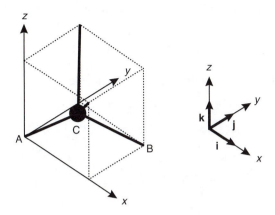

Figure 6.11   A carbon atom in a tetrahedral environment.

Worked example 6.7

The methane molecule may be thought of as containing a carbon atom at the centre of a cube together with four hydrogen atoms at alternate corners. Use this information to calculate the bond angle in methane.

Chemical background

Using the Valence Shell Electron Pair Repulsion (VSEPR) model, we would expect the four regions of electron density around the central carbon atom to be as far away from each other as possible. This leads to the arrangement described in the question, which is a regular tetrahedron. Such a structure has been confirmed experimentally by means of electron diffraction.

The tetrahedral structure of methane was also predicted by van't Hoff in 1874 on the basis of the observation that only one isomer of compounds such as CH_3Cl and CH_3Br has ever been observed.

Solution to worked example 6.7

It is necessary to formulate this problem in terms of unit vectors. If we set up the cube as shown in Figure 6.11, we can define three axes x, y, and z as shown. The carbon atom C is placed at the centre, and the four hydrogen atoms at alternate corners. We choose to label two of these A and B and will be calculating the angle A–C–B. Atom A is chosen to be at the origin of our coordinate system, and we will assign the cube an arbitrary length d. We are going to determine the angle by calculating the scalar product between vectors **AC** and **BC.**

We can define each of these vectors in terms of the unit vectors **i**, **j**, and **k** by referring to Figure 6.11. One way of moving from A to C is to move along half the side of the cube in the x-direction, then by the same amount in the y-direction, and finally by the same amount again in the z-direction. This allows us to define the vector from A to C as

$$\mathbf{AC} = \tfrac{1}{2}d\mathbf{i} + \tfrac{1}{2}d\mathbf{j} + \tfrac{1}{2}d\mathbf{k}$$

To move from B to C, we need to move ½d in the negative x-direction, ½d in the negative y-direction, and ½d in the positive z-direction. We therefore have

$$\mathbf{BC} = -\tfrac{1}{2}d\mathbf{i} - \tfrac{1}{2}d\mathbf{j} + \tfrac{1}{2}d\mathbf{k}$$

Both of these vectors will have the same modulus, which is

$$|\mathbf{AC}| = |\mathbf{BC}| = \sqrt{\left(\frac{1}{2}d\right)^2 + \left(\frac{1}{2}d\right)^2 + \left(\frac{1}{2}d\right)^2}$$

$$= \sqrt{\frac{d^2}{4} + \frac{d^2}{4} + \frac{d^2}{4}}$$

$$= \sqrt{\frac{3d^2}{4}}$$

$$= d\sqrt{\frac{3}{4}}$$

and so the scalar product is

$$\mathbf{AC} \cdot \mathbf{BC} = d\sqrt{\frac{3}{4}} \times d\sqrt{\frac{3}{4}} \times \cos\theta$$

$$= d^2 \times \frac{3}{4} \times \cos\theta$$

$$= \frac{3}{4}d^2 \cos\theta$$

where θ is the required bond angle.

Calculation of the scalar product by combining the unit vectors and using

$$\mathbf{i} \cdot \mathbf{i} = \mathbf{j} \cdot \mathbf{j} = \mathbf{k} \cdot \mathbf{k} = 1$$

and

$$\mathbf{i} \cdot \mathbf{j} = \mathbf{j} \cdot \mathbf{k} = \mathbf{k} \cdot \mathbf{i} = 0$$

gives

$$\mathbf{AC} \cdot \mathbf{BC} = \left(\frac{1}{2}d\right)\left(-\frac{1}{2}d\right) + \left(\frac{1}{2}d\right)\left(-\frac{1}{2}d\right) + \left(\frac{1}{2}d\right)\left(\frac{1}{2}d\right)$$

$$= -\frac{1}{4}d^2 - \frac{1}{4}d^2 + \frac{1}{4}d^2$$

$$= -\frac{1}{4}d^2$$

These two expressions for the scalar product can now be equated to give

$$\frac{3}{4}d^2\cos\theta = -\frac{1}{4}d^2$$

Dividing both sides by d^2 and multiplying both sides by 4 gives

$$3\cos\theta = -1$$

$$\cos\theta = -\frac{1}{3}$$

and so θ can be found by taking the inverse function:

$$\theta = \cos^{-1}\left(-\frac{1}{3}\right)$$

Using a calculator set to "degrees" mode gives $\theta = 109.5°$, which is the tetrahedral value expected.

Use the calculator

buttons followed

by ... to perform this calculation.

Worked example 6.8

A unit cell can be defined in terms of the three vectors **a**, **b**, and **c** along its edges. The angle between **b** and **c** is known as α, that between **a** and **c** as β, and that between **a** and **b** as γ. A monoclinic crystal is defined as one having α and γ equal to 90° with β having a value greater than 90°. In this case, the vectors **a**, **b**, and **c** are related to the unit vectors **i**, **j**, and **k** by the equations

$$\mathbf{a} = a\sin\beta\mathbf{i} + a\cos\beta\mathbf{k}$$

$$\mathbf{b} = b\mathbf{j}$$

$$\mathbf{c} = c\mathbf{k}$$

where a, b, and c are the magnitudes of the vectors **a**, **b**, and **c**, respectively. Use the fact that the volume V of the unit cell is given by the triple scalar product

$$V = \mathbf{a}\cdot(\mathbf{b}\times\mathbf{c})$$

to obtain an expression for the volume of the unit cell in terms of a, b, c, and β. Note that, strictly speaking, we need to take the modulus $|\mathbf{a}\cdot(\mathbf{b}\times\mathbf{c})|$ of the triple product to account for the fact that it can be negative.

Chemical background

Unit cells can be characterized into various types on the basis of the relationships between the dimensions a, b, c, α, β, and γ. The most general case in which these quantities are all independent is known as a triclinic unit cell. In the hexagonal cell, a and b are equal, α and β are both 90°, and γ is 120°.

Solution to worked example 6.8

This question requires us to calculate a *triple scalar product*; it is called this because three vectors are involved and the result will be a scalar quantity.

We evaluate the vector product enclosed in brackets first and then use this result to calculate the scalar product.

The vector product is given by

$$\mathbf{b} \times \mathbf{c} = b\mathbf{j} \times c\mathbf{k}$$

$$= bc\mathbf{j} \times \mathbf{k}$$

because the scalar quantities b and c can be multiplied directly. Because $\mathbf{j} \times \mathbf{k} = \mathbf{i}$, we now have

$$\mathbf{b} \times \mathbf{c} = bc\mathbf{i}$$

We are now able to calculate the required scalar product, which is given by

$$\mathbf{a} \cdot (\mathbf{b} \times \mathbf{c}) = (a \sin \beta \mathbf{i} + a \cos \beta \mathbf{k}) \cdot (bc\mathbf{i})$$

$$= (a \sin \beta \mathbf{i}) \cdot (bc\mathbf{i}) + (a \cos \beta \mathbf{k}) \cdot (bc\mathbf{i})$$

$$= abc \sin \beta \mathbf{i} \cdot \mathbf{i} + abc \cos \beta \mathbf{k} \cdot \mathbf{i}$$

multiplying out the terms in the brackets. Because $\mathbf{k} \cdot \mathbf{i} = 0$ and $\mathbf{i} \cdot \mathbf{i} = 1$, this leaves

$$\mathbf{a} \cdot (\mathbf{b} \times \mathbf{c}) = abc \sin \beta$$

which is equal to the volume of the monoclinic unit cell.

6.5   X-ray diffraction

Bragg's law gives us information about the behaviour of diffracted x-rays from one particular set of planes in a crystal. However, as we have seen above, there are many planes existing in a crystal lattice. Interpretation of the complex patterns, arising from the diffraction of x-rays from many different crystal planes simultaneously, enables us to obtain detailed information on crystal structure. It is generally possible to locate the positions of atoms in a unit cell, and these positions are generally expressed in terms of fractions of each of the dimensions a, b, and c, known as fractional crystallographic coordinates and often written as x/a, y/b, and z/c.

6.5.1 Complex numbers

A complex number z is one of the form

$$z = a + ib$$

where a and b are "real" numbers, which may be positive or negative and have values such as −3.4, 2.6, 7.0, and so on. The number i is defined by the equation

$$i^2 = -1$$

which seems strange at first when you realize that taking the square root of either side leads to

$$i = \sqrt{-1}$$

This i should not be confused with the unit vector **i**, which we met earlier in this chapter. You may also see the square root of −1 represented by the symbol j, not to be confused with the unit vector **j**.

In a complex number z, the number a is known as the *real* part of z, and b is known as the *imaginary* part of z. So, if we have the complex number

$$z = 2 - 3i$$

its real part is 2 and its imaginary part −3.

In chemistry, such numbers have uses in crystallography and also in quantum mechanics, as we will see in Chapter 7.

We can perform the usual arithmetic operations on complex numbers. In the case of addition and subtraction, we need to treat the real and imaginary parts separately, so that for example,

$$(3 - 2i) + (5 + i) = (3 + 5) + (-2i + i)$$

$$= 8 - i$$

In the case of multiplication, we need to consider every possible term as we have seen previously when multiplying a pair of brackets. For example,

$$(2 - 3i)(4 + i) = 2 \times (4 + i) - 3i \times (4 + i)$$

$$= 8 + 2i - 12i - 3i^2$$

Because $i^2 = -1$, this becomes

$$8 - 10i - 3(-1) = 8 - 10i + 3$$

$$= 11 - 10i$$

There is a useful relationship regarding the exponential of an imaginary number, one which contains only an imaginary part. This is

$$e^{ikx} = \cos kx + i \sin kx$$

$$e^{-ikx} = \cos kx - i \sin kx$$

Worked example 6.9

The structure factor $F(h\,k\,l)$ for a particular reflection of x-rays from a crystal is made of contributions from each of the atoms $(x_j,\,y_j,\,z_j)$ in a unit cell and is governed by the values of the integers h, k, and l. The expression for its calculation is

$$F(h\,k\,l) = \sum f_j \exp 2\pi i(hx_j + ky_j + lz_j)$$

where f_j is the atomic scattering factor for atom j. In a face-centred cubic lattice, the atomic positions $(x_j,\,y_j,\,z_j)$ are $(0, 0, 0)$, $(\frac{1}{2},\frac{1}{2}, 0)$, $(\frac{1}{2}, 0, \frac{1}{2})$, and $(0, \frac{1}{2}, \frac{1}{2})$. Show that if h, k, and l are all even or all odd, $F(h\,k\,l) = 4f_j$, but otherwise $F(h\,k\,l) = 0$.

Chemical background

The structure factor gives the amplitude of all the diffracted waves when the contributions from each atom in the unit cell are added together. The square of this is then proportional to the intensity that can be measured. The atomic scattering factor is a measure of how strongly an atom diffracts and, because the electrons are responsible for diffraction, it is proportional to the number of electrons in an atom of a given type.

The integers h, k, and l define particular planes in the crystal and, as this problem shows, certain combinations of these values lead to zero structure factors and hence intensities for certain symmetries. In fact, these systematic absences, as they are known, are characteristic of particular space groups.

International Tables for Crystallography lists the reflection conditions for the space group P2$_1$/c as

$h\,0\,l\;l = 2n$

$0\,k\,0\;k = 2n$

$0\,0\,l\;l = 2n$

Thus, for odd values of the indices specified, no reflections will be seen.

Solution to worked example 6.9

The first stage is to expand the expression given for $F(h\,k\,l)$ in terms of the cos and sin functions. It is worth remembering that $\exp(ax)$ means e^{ax}, so these are exponential functions. Consequently,

$$F(h\,k\,l) = \sum f_j \exp(2\pi i(hx_j + ky_j + lz_j))$$

$$= \sum f_j \left[\cos 2\pi(hx_j + ky_j + lz_j) + i \sin 2\pi(hx_j + ky_j + lz_j) \right]$$

Remembering that the symbol Σ denotes summation, we now need to write this expression out in full by substituting the values of $(x_j,\,y_j,\,z_j)$ for each of the four positions given. We do this and obtain

$$F(h\,k\,l) = f_1\left[\cos 0 + i\sin 0\right]$$

$$+ f_2\left[\cos 2\pi\left(\frac{1}{2}h + \frac{1}{2}k\right) + i\sin 2\pi\left(\frac{1}{2}h + \frac{1}{2}k\right)\right]$$

$$+ f_3\left[\cos 2\pi\left(\frac{1}{2}h + \frac{1}{2}l\right) + i\sin 2\pi\left(\frac{1}{2}h + \frac{1}{2}l\right)\right]$$

$$+ f_4\left[\cos 2\pi\left(\frac{1}{2}k + \frac{1}{2}l\right) + i\sin 2\pi\left(\frac{1}{2}k + \frac{1}{2}l\right)\right]$$

This can be simplified since, for example,

$$\cos 2\pi\left(\frac{1}{2}h + \frac{1}{2}k\right) = \cos \pi(h + k)$$

and we know that $\cos 0° = 1$ and $\sin 0° = 0$, so we obtain

$$F(h\,k\,l) = f_1 + f_2\left[\cos \pi(h+k) + i\sin \pi(h+k)\right]$$

$$+ f_3\left[\cos \pi(h+l) + i\sin \pi(h+l)\right]$$

$$+ f_4\left[\cos \pi(k+l) + i\sin \pi(k+l)\right]$$

If we look at Figure 6.5(a), we see that the value of $\sin \theta$ is zero when $\theta = -2\pi, -\pi, 0, \pi,$ and 2π. These are multiples of π, so all of the sine terms in the preceding equation will be zero because $h, k,$ and l are all integers. If all atoms are of the same type, we put $f_j = f$, and so this leaves us with

$$F(h\,k\,l) = f[1 + \cos \pi (h + k) + \cos \pi (h + l) + \cos \pi(k + l)]$$

If we now look at Figure 6.5(b), we see that $\cos \theta$ is 1 when θ has the values $-2\pi, 0,$ and 2π: in other words, at multiples of 2π. When θ has values of $-\pi$ and π, it is equal to -1; in other words, at odd multiples of π.

We have four cases to consider. First, if $h, k,$ and l are all even, then the quantities $\pi(h + k), \pi(h + l),$ and $\pi(k + l)$ will all be even multiples of π, and so we have

$$F(h\,k\,l) = f[1 + 1 + 1 + 1] = 4f$$

Equally, if $h, k,$ and l are all odd, then the quantities $\pi(h + k), \pi(h + l),$ and $\pi(k + l)$ will all be even multiples of π, and so, as in the preceding equation,

$$F(h\,k\,l) = 4f$$

The situation is more complicated if one or two of h, k, and l are odd. For example, if h is odd and k and l are both even, then we have

$$h + k \text{ is odd}$$

$$h + l \text{ is odd}$$

$$k + l \text{ is even}$$

Because odd multiples of π give $\cos \theta = -1$, and even multiples of π give $\cos \theta = 1$, this will give

$$F(h\ k\ l) = f\,[1 - 1 - 1 + 1] = 0$$

The same reasoning applies if k or l is odd and the other two indices are even. Now suppose two of h, k, and l are odd. If both h and k are odd, we have

$$h + k \text{ is even}$$

$$h + l \text{ is odd}$$

$$k + l \text{ is odd}$$

which leads to

$$F(h\ k\ l) = f\,[1 + 1 - 1 - 1] = 0$$

The same applies for other pairs of odd values of h, k, and l.

6.6   Symmetry operators

There are several types of symmetry elements, which include mirror planes, rotation axes, and centres of symmetry.

We have already met the idea that crystals consist of regular repeating units. Another way of expressing this is to say that they possess symmetry. As well as being important in crystals, symmetry may also be evident at the molecular level. For example, when we considered methane in worked example 6.4 it was obvious that all the hydrogen atoms were equivalent and that we could choose any pair when calculating the bond angle. In fact, if you were to rotate the methane molecule by 120° about any bond, then you would obtain a molecule that was indistinguishable in orientation from the one you started with.

The subject of symmetry is quite complex, and a thorough understanding requires some knowledge of the branch of mathematics known as group theory. Here we will only look at how individual symmetry operations can be represented.

6.6.1   Matrices

A matrix is a set of numbers arranged in rows and columns and enclosed by one pair of brackets per matrix. An example would be

$$\begin{pmatrix} 2 & 3 \\ 4 & 5 \end{pmatrix}$$

This is known as a 2×2 matrix because it has 2 rows and 2 columns. A 2×3 matrix has 2 rows and 3 columns, and an example would be

$$\begin{pmatrix} 1 & 2 & 3 \\ 7 & 8 & 9 \end{pmatrix}$$

We can only calculate the sum or difference of two matrices if they have the same dimensions. The operation is performed using the corresponding elements of the two matrices. For example, if

$$\mathbf{A} = \begin{pmatrix} 2 & 3 \\ 4 & 5 \end{pmatrix} \quad \text{and} \quad \mathbf{B} = \begin{pmatrix} 1 & 2 \\ 3 & 4 \end{pmatrix}$$

we have

$$\mathbf{A} + \mathbf{B} = \begin{pmatrix} 2 & 3 \\ 4 & 5 \end{pmatrix} + \begin{pmatrix} 1 & 2 \\ 3 & 4 \end{pmatrix}$$

$$= \begin{pmatrix} 2+1 & 3+2 \\ 4+3 & 5+4 \end{pmatrix}$$

$$= \begin{pmatrix} 3 & 5 \\ 7 & 9 \end{pmatrix}$$

and

$$\mathbf{A} - \mathbf{B} = \begin{pmatrix} 2 & 3 \\ 4 & 5 \end{pmatrix} - \begin{pmatrix} 1 & 2 \\ 3 & 4 \end{pmatrix}$$

$$= \begin{pmatrix} 2-1 & 3-2 \\ 4-3 & 5-4 \end{pmatrix}$$

$$= \begin{pmatrix} 1 & 1 \\ 1 & 1 \end{pmatrix}$$

It is also possible to multiply a matrix by a number, which effectively acts as a scaling factor. For the example above, we could scale matrix $\mathbf{A}$ by a factor of 2 so that

$$2\mathbf{A} = 2\begin{pmatrix} 2 & 3 \\ 4 & 5 \end{pmatrix} = \begin{pmatrix} 2 \times 2 & 2 \times 3 \\ 2 \times 4 & 2 \times 5 \end{pmatrix} = \begin{pmatrix} 4 & 6 \\ 8 & 10 \end{pmatrix}$$

Two matrices can only be multiplied together if the number of columns in the first matrix is the same as the number of rows in the second. It is possible for a 3×2 matrix to be multiplied by a 2×1 matrix; the result would be a 3×1 matrix. However, you cannot multiply a 2×1 matrix by a 3×2 matrix because the number of columns in the first (1) does not match the number of rows in the second (3). The order of multiplying matrices is therefore important.

To perform the multiplication, we need to multiply each element in the rows of the first matrix by the corresponding elements in the columns of the second matrix. These pairs of numbers are then added to give the element of the matrix, which appears in the position given by the intersection of this row and column. For example, if we wish to calculate

$$\begin{pmatrix} 2 & 4 \\ 6 & 8 \\ 7 & 1 \end{pmatrix}\begin{pmatrix} 1 & 2 \\ 5 & 6 \end{pmatrix}$$

then, because we are multiplying a 3×2 matrix by a 2×2 matrix, we expect to obtain a 3×2 matrix. Matching elements from the first row of the first matrix and the first column of the second matrix gives us

$$(2 \times 1) + (4 \times 5) = 22$$

This result appears in the position defined by the intersection, i.e., the *first* row and the *first* column in the results matrix. Now, working along the *second* row of the first matrix and the *first* column of the second matrix, we have

$$(6 \times 1) + (8 \times 5) = 46$$

and this appears in the position defined by the first column and the second row of the results matrix. We build up the whole matrix in this way to give us the final result:

$$\begin{pmatrix} (2 \times 1) + (4 \times 5) & (2 \times 2) + (4 \times 6) \\ (6 \times 1) + (8 \times 5) & (6 \times 2) + (8 \times 6) \\ (7 \times 1) + (1 \times 5) & (7 \times 2) + (1 \times 6) \end{pmatrix} = \begin{pmatrix} 22 & 28 \\ 46 & 60 \\ 12 & 20 \end{pmatrix}$$

It is possible to define a matrix for any symmetry operation. For example, the centre of symmetry converts a set of coordinates (x, y, z) into $(-x, -y, -z)$ and could be represented by the matrix

$$\begin{pmatrix} -1 & 0 & 0 \\ 0 & -1 & 0 \\ 0 & 0 & -1 \end{pmatrix}$$

Worked example 6.10

The operations defined by the symmetry elements in methane can be expressed in matrix notation. The four hydrogen atoms can be represented by the 1×4 matrix

$$(H_A \ H_B \ H_C \ H_D)$$

Multiply this matrix by each of the following 4 × 4 symmetry matrices to determine the 1 × 4 matrix, which defines the resulting position of each hydrogen atom:

(a)
$$\begin{pmatrix} 0 & 1 & 0 & 0 \\ 0 & 0 & 1 & 0 \\ 1 & 0 & 0 & 0 \\ 0 & 0 & 0 & 1 \end{pmatrix}$$

(b)
$$\begin{pmatrix} 0 & 0 & 0 & 1 \\ 1 & 0 & 0 & 0 \\ 0 & 1 & 0 & 0 \\ 0 & 0 & 1 & 0 \end{pmatrix}$$

(c)
$$\begin{pmatrix} 0 & 0 & 1 & 0 \\ 0 & 0 & 0 & 1 \\ 0 & 1 & 0 & 0 \\ 1 & 0 & 0 & 0 \end{pmatrix}$$

(d)
$$\begin{pmatrix} 0 & 0 & 1 & 0 \\ 0 & 1 & 0 & 0 \\ 0 & 0 & 0 & 1 \\ 1 & 0 & 0 & 0 \end{pmatrix}$$

Chemical background

Methane is said to belong to the point group known as T_d. A point group characterizes the symmetry elements possessed by a molecule; in the case of T_d, these are listed as E, $8C_3$, $3C_2$, $6S_4$, and $6\sigma_d$. E is an identity operation that leaves the molecule unchanged. C_3 represents a rotation of 360°/3 (i.e., 120°), and C_2 a rotation of 360°/2 (i.e., 180°). S_4 is known as an improper rotation and involves rotation of 360°/4 (i.e., 90°), followed by reflection across the symmetry plane perpendicular to the axis of rotation. σ_d is a dihedral symmetry plane that bisects the angle formed by a pair of C_2 axes.

Solution to worked example 6.10

Application of the first symmetry operator matrix requires us to perform the multiplication

$$\begin{pmatrix} H_A & H_B & H_C & H_D \end{pmatrix} \begin{pmatrix} 0 & 1 & 0 & 0 \\ 0 & 0 & 1 & 0 \\ 1 & 0 & 0 & 0 \\ 0 & 0 & 0 & 1 \end{pmatrix}$$

This is a 1 × 4 matrix multiplied by a 4 × 4 matrix, and will produce a 1 × 4 matrix result.

Figure 6.12   A three-fold rotation applied to a methane molecule.

We calculate this as shown in the example above, but the presence of several zeros in the symmetry operator matrix assists this process considerably. We will generate the 1×4 matrix by matching elements in the initial 1×4 matrix with elements in each successive column of the 4×4 matrix. For the first column, we obtain

$$(H_A \times 0) + (H_B \times 0) + (H_C \times 1) + (H_D \times 0) = H_C$$

Subsequent columns give

$$(H_A \times 1) + (H_B \times 0) + (H_C \times 0) + (H_D \times 0) = H_A$$

$$(H_A \times 0) + (H_B \times 1) + (H_C \times 0) + (H_D \times 0) = H_B$$

$$(H_A \times 0) + (H_B \times 0) + (H_C \times 0) + (H_D \times 1) = H_D$$

so that the final matrix is

$$(H_C\ H_A\ H_B\ H_D)$$

Comparison with the initial matrix shows that H_A has moved to the position formerly occupied by H_B, H_B to that by H_C, H_C to that by H_A, and H_D has remained in the same position. The matrix consequently represents a rotation of $120°$ about the $C–H_D$ bond, and the net result of performing this symmetry operation is shown in Figure 6.12.

Similar calculations give the following results for applications of the remaining symmetry operator matrices:

(b) $(H_B\ H_C\ H_D\ H_A)$

(c) $(H_D\ H_C\ H_A\ H_B)$

(d) $(H_D\ H_B\ H_A\ H_C)$

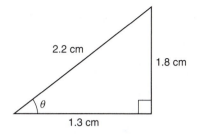

Figure 6.13   The right-angled triangle referred to in question 4.

Exercises

1. The two perpendicular sides in a right-angled triangle have the lengths 4.62 and 8.31 cm. What is the length of the hypotenuse?

2. Calculate the distance between the following pairs of points:

    (a)   (1,2) and (3,4)

    (b)   (7,–2) and (3,1)

    (c)   (–6,–5) and (–3,2)

3. Calculate the distance between the following pairs of points:

    (a)   (3,1,6) and (4,3,4)

    (b)   (2,1,–3) and (1,0,–1)

    (c)   (5,–9,–2) and (6,–1,3)

4. In the right-angled triangle shown in Figure 6.13, calculate $\sin \theta$, $\cos \theta$, and $\tan \theta$, giving your answers correct to 2 decimal places.

5. Determine the angles in the triangle defined in Figure 6.13.

6. Calculate the following, giving your answers correct to 3 decimal places:

    (a) $\sin 11°$

    (b) $\cos(-32°)$

    (c) $\tan 108°$

7. Calculate the following to 4 decimal places:

    (a) $\sin(\pi/3)$

    (b) $\cos(-3\pi/4)$

    (c) $\tan(\pi/6)$

8. Solve the following equations, giving your answer in degrees.

(a) $\sin x = 0.352$

(b) $\cos x = -0.108$

(c) $\tan x = 0.764$

9. Solve the following equations, giving your answer in radians.

(a) $\tan(2x + 1) = 0.472$

(b) $\cos(3x + 2) = 0.564$

(c) $\sin(x - 1) = 0.552$

10. Calculate the modulus of these vectors:

(a) $2\mathbf{i} - 2\mathbf{j} - 3\mathbf{k}$

(b) $2\mathbf{i} + 4\mathbf{j} - 6\mathbf{k}$

(c) $-3\mathbf{i} + 2\mathbf{j} + 5\mathbf{k}$

11. Determine a unit vector in the direction of each of the following vectors:

(a) $4\mathbf{i} - 2\mathbf{j} + 6\mathbf{k}$

(b) $-\mathbf{i} - 2\mathbf{j} + 7\mathbf{k}$

(c) $2\mathbf{i} - 6\mathbf{j} + 2\mathbf{k}$

12. If $\mathbf{a} = \mathbf{i} + 2\mathbf{j} - 3\mathbf{k}$ and $\mathbf{b} = 2\mathbf{i} - \mathbf{j} + \mathbf{k},$ calculate

(a) $\mathbf{a} + \mathbf{b}$

(b) $\mathbf{a} - \mathbf{b}$

(c) $\mathbf{a} \cdot \mathbf{b}$

(d) $\mathbf{a} \times \mathbf{b}$

13. Determine the angle between the vectors $2\mathbf{i} + 2\mathbf{j} + \mathbf{k}$, and $\mathbf{i} - \mathbf{j} - 2\mathbf{k}$.

14. Determine the real and imaginary parts of each of the following complex numbers:

(a) $4 + 2i$

(b) $1 - 8i$

(c) $3 - 2i$

15. Find the values of the following expressions involving complex numbers:

   (a) $(3 + 6i) + (4 - 3i)$

   (b) $(1 + 6i) + (7 - 4i)$

   (c) $(9 - 8i) - (7 + 4i)$

16. Determine the following products of complex numbers:

   (a) $(5 + 2i)(3 - i)$

   (b) $(1 + i)(2 + 3i)$

   (c) $(4 - 6i)(3 + 2i)$

17. Obtain expressions for $\cos kx$ and $\sin kx$ in terms of e^{ikx} and e^{-ikx}.

18. If

$$A = \begin{pmatrix} 2 & 2 \\ 3 & 1 \end{pmatrix} \quad \text{and} \quad B = \begin{pmatrix} 1 & 0 \\ 4 & 2 \end{pmatrix}$$

   calculate

   (a) $A + B$

   (b) $A - B$

   (c) AB

19. Calculate the product

$$\begin{pmatrix} 3 & 0 & 1 \\ 2 & 4 & 2 \\ 1 & 1 & 3 \end{pmatrix} \begin{pmatrix} 2 \\ 1 \\ 1 \end{pmatrix}$$

20. If

$$A = \begin{pmatrix} 1 & 2 \\ 4 & 2 \end{pmatrix} \quad \text{and} \quad B = \begin{pmatrix} 2 & 3 \\ 3 & 1 \end{pmatrix}$$

   calculate $(2A + B)A$.

Problems

1. In a simple lattice the volume of atoms is $1 \times \dfrac{4}{3}\pi r^3$, and the volume of the unit cell is $(2r)^3$. What is the value of the packing fraction?

2. Two carbon atoms in a layer of graphite have coordinates (1.24, 2.62) and (2.09, 3.82) with the distances measured in Ångströms. What is the distance between this pair of atoms?

3. A metal complex has a copper atom at (6.562, 10.888, 4.923) and an oxygen atom bound to it at (8.290, 10.352, 4.275). What is the Cu–O bond length if these measurements in Ångströms?

4. The second-order reflections from a potassium chloride crystal occur at an angle of 29.2° when radiation of wavelength 1.537 Å is used. Calculate the spacing of the planes responsible for this diffraction.

5. The (111) planes in molybdenum carbide have a spacing of 2.47×10^{-10} m. What is the first-order diffraction angle when x-rays of wavelength 2.94×10^{-10} m are fired at this crystal?

6. Suppose that the atoms A and B in Figure 6.11 are not equivalent so that the bond lengths AC and BC are different and

$$\mathbf{AC} = 0.5\, d\, \mathbf{i} + 0.5\, d\, \mathbf{j} + 0.4\, d\, \mathbf{k}$$

$$\mathbf{BC} = -0.5\, d\, \mathbf{i} - 0.5\, d\, \mathbf{j} + 0.6\, d\, \mathbf{k}$$

Calculate the new value of the bond angle.

7. The atoms in a body-centred cubic lattice are situated at the corners of a cube and at its centre, as shown in Figure 6.14. What are the four unique interatomic vectors in this lattice if the cube has side a?

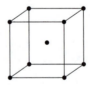

Figure 6.14   A body-centred cubic lattice.

8. Calculate the volume of the unit cell in the mineral malachite, which has the formula $Cu_2(OH)_2CO_3$ and crystallizes in the space group $P2_1/a$. The unit cell is monoclinic and has dimensions $a = 9.502$, $b = 11.974$, $c = 3.240$ Å, and $\beta = 98.75°$.

9. A crystal containing a twofold screw axis has equivalent positions (x,y,z) and $(-x, -y, z + 0.5)$.

   (a) Obtain a general expression for $F(0\ 0\ l)$.
   (b) By considering the positions (0,0,0) and (0,0,0.5) show that $F(0\ 0\ l) \neq 0$ when $l = 2n$.

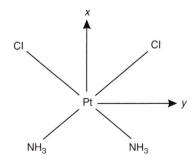

Figure 6.15   The structure of cisplatin.

10. The anticancer agent cisplatin has the structure shown in Figure 6.15, which also shows a set of coordinate axes that can be used to define the positions of each atom.
Write a matrix equation for finding the 2×1 matrix

$$\begin{pmatrix} x_2 \\ y_2 \end{pmatrix}$$

which defines the position of N2 or Cl2 starting from the corresponding matrix containing x_1 and y_1, which defines the position of the other atom of the same type (N1 and Cl1).

Quantum mechanics 7

The ideas involved in quantum mechanics can be expressed quite concisely, yet a detailed study of individual quantum mechanical systems seems to rely heavily upon mathematics. In some ways this is unfortunate, as there is a danger that the underlying concepts may be lost in this detail. However, it is essential to be able to apply these ideas if we are to obtain results that are of use, and which may be compared with experimental data. In this chapter, we will deal with the mathematics needed for some of the simpler concepts before looking at more complicated situations that need more specialized tools.

7.1 Energy-level transitions and appropriate precision

It is important when performing calculations to quote results to an appropriate number of figures, as discussed in Chapter 2, Section 2.2.2, on page 25. Often we are concerned not to give too many figures. However, in quantum mechanical calculations there are times when the temptation is to give too few.

Worked example 7.1

In 1963 the SI unit of length, the metre, was defined as 1 650 763.73 wavelengths of the orange-red radiation from krypton-86. Calculate the (a) wavelength, (b) frequency, and (c) energy of this radiation.

Chemical background

Krypton-86 is one of the five stable isotopes of this element. Krypton is a colourless gas that is generally inert, although it can form compounds with fluorine. It finds use in fluorescent lamps.

Isotopes are atoms of the same element that differ only in the numbers of neutrons in their nuclei.

Solution to worked example 7.1

(a) If there are 1 650 763.73 wavelengths in exactly 1 m, the length of a single wavelength will be

$$\frac{1}{1650763.73}\,\mathrm{m} = 6.05780211 \times 10^{-7}\,\mathrm{m}$$

Notice that because we are dealing with a length of exactly 1m, it is the number of significant figures in the denominator that determines the number of significant figures in the answer.

Use the x^{-1} button to determine a reciprocal value.

(b) Frequency ν is calculated from wavelength l using the formula

$$\nu = \frac{c}{\lambda}$$

where c is the velocity of light in a vacuum. In order to avoid losing precision, we need to take the value of c to at least 9 significant figures; this is $2.997\ 924\ 58 \times 10^8$ m s^{-1}. The calculation is then

$$\nu = \frac{c}{\lambda} = \frac{2.99792458 \times 10^8 \, \text{m s}^{-1}}{6.05780211 \times 10^{-7} \, \text{m}} = 4.94886516 \times 10^{14} \text{s}^{-1}$$

Alternatively, this could be written as $4.94886516 \times 10^{14}$ Hz. Note that choosing the value of c to 9 significant figures means that we can also quote ν to 9 significant figures.

(c) Energy E is given by the formula

$$E = h\nu$$

where h is the Planck constant. Because ν was obtained to 9 significant figures in the previous part, we can ensure that E can be quoted to this level of precision by choosing a value of h to 9 significant figures. However, the uncertainty in h means that it is typically quoted to a maximum of 8 significant figures, viz.,

$$h = 6.626\ 075\ 5 \times 10^{-34} \, \text{J s}$$

Using this value gives

$$E = h\nu$$

$$= 6.626\ 075\ 5 \times 10^{-34} \, \text{J s} \times 4.94886516 \times 10^{14} \, \text{s}^{-1}$$

$$= 3.279\ 155\ 4 \times 10^{-19} \, \text{J}$$

where we are only justified in quoting 8 significant figures due to the limitation in the value of h.

Use the ⬛ button on your calculator to enter powers of 10; for negative powers use the ⬛ button.

7.2   The photon

The concept of a photon is central to an appreciation of quantum mechanics. It is one unit of electromagnetic radiation, also known as a quantum. This is an example of the idea that quantities traditionally regarded as waves, in this case radiation, can behave as discrete particles.

7.2.1   Mathematical relationships

We saw earlier in Chapter 3, Section 3.7.3, on page 88 that if we know that a pair of variables are related proportionally, then it is possible to obtain an expression for the relationship between them. Sometimes, we do not know what this relationship is, or even whether it exists at all. We then need to inspect the data given to see if it is possible to obtain such a relationship.

Worked example 7.2

Derive an equation that relates the energy E of a photon to its frequency ν, given the following data:

$\dfrac{E}{10^{-19}\,\text{J}}$	$\dfrac{\nu}{10^{14}\,\text{s}^{-1}}$
19.89	30
9.945	15
6.630	10
4.973	7.5
3.978	6.0

Calculate the energy corresponding to a frequency of $8.57 \times 10^{14}\,\text{s}^{-1}$.

Chemical background

When light is absorbed by the surface of a metal, electrons are ejected, but this only happens when the frequency of light is above a certain value. Classically, we would expect light of high intensity but low frequency to also cause the ejection of electrons, and this discrepancy led Einstein to propose that light consisted of discrete photons.

This is the basis of the well-known photoelectric effect, which is often used as an introduction to quantum mechanics. The experimental arrangement for this demonstration is shown in Figure 7.1. The maximum kinetic energy of the electrons is proportional to the difference between the frequency of the incident radiation and the value known as the threshold frequency.

Figure 7.1   The apparatus used to demonstrate the photoelectric effect.

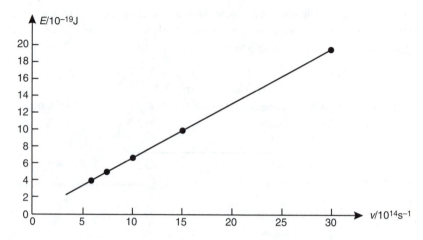

Figure 7.2   Graph of photon energy against frequency using the data given in worked example 7.2.

Solution to worked example 7.2

Notice that the table headings are given as quotients including the units, as we have seen previously. Although it would be possible to expand each value in the table in terms of its units, it is actually neater *not* to do this but to work with the table headings as they are given.

A casual inspection of the data shows that as the values of E increase, so do the values of ν, as shown in Figure 7.2. The simplest relationship we could hope to find, then, would be that of direct proportion. In that case, the ratio E/ν would be constant and equal to the constant of proportionality. If the ratio E/ν is constant, it is also true that the ratio

$$\frac{E/_{10^{-19}J}}{\nu/_{10^{14}s^{-1}}}$$

is a constant, this being easier to calculate from the data given. Our table of data can then be extended to include this calculation.

$\dfrac{E}{10^{-19}J}$	$\dfrac{\nu}{10^{14}s^{-1}}$	$\dfrac{E/_{10^{-19}J}}{\nu/_{10^{14}s^{-1}}}$
19.89	30	0.663
9.945	15	0.663
6.630	10	0.663
4.973	7.5	0.663
3.978	6.0	0.663

The constant ratio term indicates that E and ν are indeed directly proportional and are related by the equation

$$\frac{E/10^{-19}\,J}{\nu/10^{14}\,s^{-1}} = 0.663$$

We can multiply both sides by $\nu/10^{14}$ s^{-1} so that this becomes

$$\frac{E}{10^{-19}\,J} = \frac{0.663 \times \nu}{10^{14}\,s^{-1}}$$

We can also multiply both sides by 10^{-19} J to obtain

$$E = 0.663 \times \frac{\nu}{10^{14}\,s^{-1}} \times 10^{-19}\,J$$

The powers of 10, $10^{-19}/10^{14}$ combine to give $10^{(-19-14)}$ or 10^{-33}, and the unit J/s^{-1} is the same as J s. This then gives the relationship as

$$E = (0.663 \times 10^{-33}\ J\ s)\nu$$

$$= (6.63 \times 10^{-34}\ J\ s)\nu$$

The quantity in brackets is known as *Planck's constant* and is normally given the symbol h, so we obtain the relationship

$$E = h\nu$$

where $h = 6.63 \times 10^{-34}$ J s. This expression should be familiar from the previous section.

We can now substitute in the value of ν of 8.57×10^{14} s^{-1} to give

$$E = 6.63 \times 10^{-34}\ J\ s \times 8.57 \times 10^{14}\ s^{-1}$$

$$= 5.68 \times 10^{-19}\ J$$

because the units s and s^{-1} cancel. Notice that because the frequency value of 8.57×10^{14} s^{-1} falls within the range of the values given in the question ($6.0 - 30 \times 10^{14}$ s^{-1}), this is known as interpolation. If we had been calculating the energy by using a frequency value outside this range, it would be known as extrapolation (Figure 7.3). We can be less confident of extrapolating reliably because the relationship we have obtained may not be valid over other ranges of frequency.

Powers of 10 are entered using the EXP button.

Figure 7.3   Interpolation and extrapolation.

7.3   Forces between atoms

Despite the successes of quantum mechanics, the classical view of atoms can still be very useful and is more appropriate in some circumstances. We are often interested in the interactions between atoms, as these allow the prediction of properties such as the energy and structure of a system.

7.3.1   Proportionality

The subject of proportion was discussed in Chapter 3, Section 3.7.3, on page 88. There we saw that if two quantities x and y are proportional to each other, then $y/x = k$, or $y = kx$, where k is the constant of proportionality.

Worked example 7.3

One model of a chemical bond is that of a simple spring, as shown in Figure 7.4, in which the force acting on the spring is proportional to the displacement of the bond length from its equilibrium value. In this case, the constant of proportionality is known as the force constant of the bond and has the value of 440 N m^{-1} for a single bond between a pair of carbon atoms. Calculate the force on this bond if it is displaced by 2.5 pm from its equilibrium position.

Figure 7.4   The (a) displacement of a spring from its (b) equilibrium position.

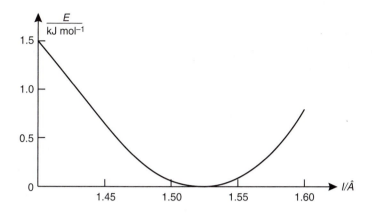

Figure 7.5   The energy arising from the displacement of a bond from its equilibrium length.

Chemical background

Such calculations form the basis of the molecular mechanics method for calculating molecular structures and energies. The energy E arising from the displacement of a particular bond is calculated from an equation of the form

$$E = \frac{1}{2}k(l - l_o)^2$$

where k is the force constant, l the observed length, and l_o the equilibrium length. Figure 7.5 shows a plot of E against l for the displacement of the bond being discussed here.

Typical values of these parameters for a single bond between a pair of carbon atoms would be $k = 4.40$ mdyn $Å^{-1}$ and $l_o = 1.523$ Å. These non-SI units are invariably used in calculations of this type.

Solution to worked example 7.3

We need to set up an equation to relate the force F to the displacement x. Because $F \propto x$, it follows that

$$F = kx$$

and we are given the value of the proportionality constant k in the question. We substitute the values given into this equation to obtain

$$F = 440 \text{ N m}^{1} \times 2.5 \text{ pm}$$

$$= 440 \text{ N m}^{-1} \times 2.5 \times 10^{-12} \text{ m}$$

$$= 1.1 \times 10^{-9} \text{ N}$$

$$= 1.1 \text{ nN}$$

7.3.2   Stationary points

Stationary points were discussed in Chapter 4, Section 4.6.2, on page 127. We saw that the stationary points of the function $f(x)$ occurred when

$$\frac{df(x)}{dx} = 0$$

and that

$$\frac{d^2 f(x)}{dx^2} < 0$$

denotes a maximum, whereas

$$\frac{d^2 f(x)}{dx^2} > 0$$

denotes a minimum.

Worked example 7.4

The energy $E(r)$ of the interaction between two molecules separated by a distance r is given by the function

$$E(r) = -\frac{A}{r^6} + \frac{B}{r^{12}}$$

The constants A and B are related to the depth ε of the potential well and the equilibrium intermolecular separation r_e. For xenon these quantities have the values 1.9 kJ mol^{-1} and 4.06 Å, respectively, the potential curve being illustrated in Figure 7.6.

Locate and identify the stationary points of the function $E(r)$.

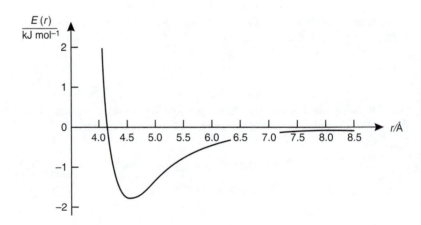

Figure 7.6   Graph of the interaction energy $E(r)$ as defined in worked example 7.4.

Chemical background

The function given represents the interaction energy between a pair of molecules, but such functions have been used as "interatomic potentials" to model the interactions of atoms in the solid phase. Other types of function can be used to represent these interactions, and parameters such as A and B need to be varied to give the best agreement with experimental data.

Solution to worked example 7.4

We saw in Chapter 4, Section 4.6.2 that the stationary points of a function $f(x)$ were given by those values of x that make

$$\frac{df(x)}{dx} = 0$$

and that the nature of those stationary points is determined by evaluating the sign of

$$\frac{d^2 f(x)}{dx^2}$$

The differentiation of $E(r)$ is easier to perform if we rewrite it as

$$E(r) = -Ar^{-6} + Br^{-12}$$

because

$$\frac{1}{x^n} = x^{-n}$$

Using the rule that

$$\frac{d}{dx}(ax^n) = anx^{n-1}$$

then gives us

$$\frac{d}{dr}(-Ar^{-6}) = (-A)(-6)r^{-6-1}$$

$$= 6Ar^{-7}$$

and

$$\frac{d}{dr}(Br^{-12}) = B \times (-12r^{-12-1})$$

$$= -12Br^{-13}$$

so that

$$\frac{dE(r)}{dr} = \frac{6A}{r^7} - \frac{12B}{r^{13}}$$

This will be zero when

$$\frac{6A}{r^7} = \frac{12B}{r^{13}}$$

This equation can be rearranged by multiplying both sides by r^{13} and dividing both sides by $6A$ to give

$$\frac{r^{13}}{r^7} = \frac{12B}{6A}$$

$$r^{13-7} = \frac{2B}{A}$$

$$r^6 = \frac{2B}{A}$$

We can finally solve this equation by taking the sixth root of either side because

$$\sqrt[6]{r^6} = r$$

In other words, the process of taking the sixth root is the inverse of raising a number to the power 6. The same relationship holds for any other power. Having done this, we finally obtain

$$r = \sqrt[6]{\frac{2B}{A}}$$

This tells us the position of the stationary point but nothing about its nature. We now need to calculate $\frac{d^2E(r)}{dr^2}$, which we can do by differentiating the equation

$$\frac{dE(r)}{dr} = \frac{6A}{r^7} - \frac{12B}{r^{13}}$$

which is easier to do if we rewrite it as

$$\frac{dE(r)}{dr} = 6Ar^{-7} - 12Br^{-13}$$

Using the same rules of differentiation as before gives

$$\frac{d^2E}{dr^2} = (6 \times (-7))Ar^{-7-1} + ((-12) \times (-13))Br^{-13-1}$$

$$= -42Ar^{-8} + 156Br^{-14}$$

$$= -\frac{42A}{r^8} + \frac{156B}{r^{14}}$$

Because the minimum occurs when

$$r = \sqrt[6]{\frac{2B}{A}}$$

it follows that

$$r^6 = \frac{2B}{A}$$

It helps if we can rewrite our expression for $\frac{d^2E}{dr^2}$ in terms of powers of r^6.

$$\frac{d^2E}{dr^2} = -\frac{42A}{r^8} + \frac{156B}{r^{14}}$$

$$= \frac{1}{r^2}\left(-\frac{42A}{r^6} + \frac{156B}{r^{12}}\right)$$

Now,

$$r^{12} = (r^6)^2$$

$$= \left(\frac{2B}{A}\right)^2$$

$$= \frac{2^2 B^2}{A^2}$$

$$= \frac{4B^2}{A^2}$$

and so

$$\frac{d^2E}{dr^2} = \frac{1}{r^2}\left(-\frac{42A}{2B/A} + \frac{156B}{4B^2/A^2}\right)$$

$$= \frac{1}{r^2}\left(-\frac{42A^2}{2B} + \frac{156BA^2}{4B^2}\right)$$

$$= \frac{1}{r^2}\left(-\frac{21A^2}{B} + \frac{39A^2}{B}\right)$$

$$= \frac{1}{r^2}\frac{18A^2}{B}$$

Because A and B are both positive constants, $\dfrac{d^2E}{dr^2}$ must be greater than zero, indicating a minimum.

7.4   Particle in a box

One of the first quantum mechanical systems treated in a study of the subject is usually the particle in a box. This consists of a particle constrained within a one-dimensional box of a specified width, within which the potential energy is zero and outside of which the potential energy is infinite. The model is readily extended to two- and three-dimensional systems.

7.4.1   Complex numbers

We met the concept of complex numbers in Chapter 6, Section 6.5.1, on page 213. In particular, we saw that a complex exponential could be expressed by the equation

$$e^{ikx} = \cos kx + i \sin kx$$

Worked example 7.5

The wavefunctions Ψ for the particle in a one-dimensional box are given by the expression

$$\Psi = \left(\frac{2}{a}\right)^{\frac{1}{2}}\left(\frac{1}{2i}\right)(e^{ikx} - e^{-ikx})$$

where $i^2 = -1$, a is the length of the box, x is the position of the particle within the box, and k is a constant. Simplify this expression.

Chemical background

The constant k is actually given by the formula

$$k = \frac{\sqrt{8\pi^2 mE}}{h}$$

where m is the mass of the particle, E is its energy, and h is Planck's constant. If we apply the boundary conditions that the wavefunction must fall to zero at $x = 0$ and $x = a$, we can solve this expression to give the allowed values of the energy, as we will see in the next worked example.

Solution to worked example 7.5

We need to use the relationships

$$e^{ikx} = \cos kx + i \sin kx$$

and

$$e^{-ikx} = \cos kx - i \sin kx$$

We then have

$$\Psi = \left(\frac{2}{a}\right)^{\frac{1}{2}} \left(\frac{1}{2i}\right) (e^{ikx} - e^{-ikx})$$

$$= \left(\frac{2}{a}\right)^{\frac{1}{2}} \left(\frac{1}{2i}\right) [\cos kx + i \sin kx - (\cos kx - i \sin kx)]$$

$$= \left(\frac{2}{a}\right)^{\frac{1}{2}} \left(\frac{1}{2i}\right) [\cos kx + i \sin kx - \cos kx + i \sin kx]$$

The two terms in $\cos kx$ cancel out, and we are left with

$$\Psi = \left(\frac{2}{a}\right)^{\frac{1}{2}} \left(\frac{1}{2i}\right) [2i \sin kx]$$

$$= \left(\frac{2}{a}\right)^{\frac{1}{2}} \sin kx$$

The graph of this wavefunction is shown in Figure 7.7.

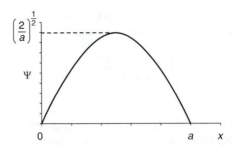

Figure 7.7   Graph of the wavefunction Ψ for the particle in a one-dimensional box of length a.

7.4.2   Sequences

In Chapter 3, Section 3.4.3, on page 64, we met the idea of a function for the first time. This was represented by $f(x)$, where x was a number upon which certain operations (such as multiplication by a constant, raising to a power, and so on) were performed. A sequence differs from a function only in that x is replaced by n, where n is an integer (i.e., a whole number) and is therefore restricted to the values 0, 1, 2, 3, 4, ...

For example, we could define a sequence as

$$f(n) = 3n + 1 \text{ for } n \geq 0$$

We would then have

$$f(0) = (3 \times 0) + 1 = 1$$

$$f(1) = (3 \times 1) + 1 = 4$$

$$f(2) = (3 \times 2) + 1 = 7$$

$$f(3) = (3 \times 3) + 1 = 10$$

and so on. It is necessary to specify the starting value of n, which will usually be 0 or 1. This can be done using a greater than sign (>) or a greater than or equal to sign ($\geq$).

In quantum mechanics, we are often concerned with the idea that particles may only occupy discrete energy levels that are also labeled with integers (usually from 1 upwards).

Worked example 7.6

The energy E of each level n in the one-dimensional particle in a box problem is given by the function

$$E(n) = \frac{n^2 h^2}{8ma^2}$$

for $n \geq 1$ where m is the mass of the particle, a is the width of the box, and h is Planck's constant. Find the sum of the first five terms of this sequence in terms of a, h, and m.

Chemical background

The pi electrons in hexatriene are delocalized, and each can be thought of as being contained in a one-dimensional box of length 7.3 Å, this being the distance between the terminal carbon atoms plus half a bond length on either side.

Solution to worked example 7.6

We need to add together five terms of the form

$$\frac{n^2 h^2}{8ma^2}$$

Once we realize that the quantity $\dfrac{h^2}{8ma^2}$ is common to each of these terms, we can write the required sum as

$$\frac{h^2}{8ma^2}(1^2 + 2^2 + 3^2 + 4^2 + 5^2) = \frac{h^2}{8ma^2}(1 + 4 + 9 + 16 + 25)$$

$$= \frac{55h^2}{8ma^2}$$

These energy levels are shown in Figure 7.8.

This is analogous to summing the energies of electrons in a molecule where each pair occupies a separate energy level.

Figure 7.8   The energy levels available to a particle confined to a one-dimensional box.

7.4.3   Inverse functions

This topic has been dealt with in Chapter 5, Section 5.6.2, on page 173. We saw that the effect of an inverse function $\text{arc}f(x)$ is to undo the effect of the original function $f(x)$. In particular,

$$f(\text{arc}f(x)) = x \text{ and } \text{arc}f(f(x)) = x$$

Worked example 7.7

The quantum mechanical energy E of a particle in a three-dimensional box is given by the function

$$E(n_x, n_y, n_z) = \frac{h^2}{8m}\left(\frac{n_x^2}{a^2} + \frac{n_y^2}{b^2} + \frac{n_z^2}{c^2}\right)$$

Because we are now dealing with three quantum numbers, it is possible to obtain degenerate energy levels. These have the same energies but different quantum numbers, for example, $E(1,2,1)$ and $E(1,1,2)$ when $a = b = c$.

where h is Planck's constant and m the mass of the particle. The quantum numbers n_x, n_y, and n_z relate to the directions x, y, and z in which the box has dimensions a, b, and c, respectively.

Obtain an expression for the inverse function $\text{arc}E(n)$ in the special case where $n_x = n_y = n_z = n$ and $a = b = c = d$ which is the situation existing in a cube.

Chemical background

This model could be used to represent, say, nitrogen gas in a 10 cm × 10 cm × 10 cm container. The calculation would then show that the separation between the energy levels is extremely small and so there is no effective quantization on this scale.

The extension of the quantum mechanics of the particle in a one-dimensional box to three dimensions is relatively straightforward. This is because the motion in the three perpendicular directions x, y, and z can be treated independently. If this was not so, the solution to the problem would be much more difficult to obtain. This ability to partition the energy has important implications in both quantum mechanics and statistical mechanics.

Solution to worked example 7.7

The first stage in this problem is to obtain an expression for $E(n)$, the energy of a particle in a cube. If we make the substitutions

$$n_x = n_y = n_z = n$$

and

$$a = b = c = d$$

into the given equation, we obtain

$$E(n,n,n) = \frac{h^2}{8m}\left(\frac{n^2}{d^2} + \frac{n^2}{d^2} + \frac{n^2}{d^2}\right)$$

$$= \frac{3h^2n^2}{8md^2}$$

Before we consider the inverse of this function, arc $E(n)$, it is worth considering the stages involved in calculating $E(n)$ from a given value of n:

1. Square n to give n^2.
2. Multiply n^2 by $3h^2$ to give $3h^2n^2$.
3. Divide $3h^2n^2$ by $8md^2$ to give

$$\frac{3h^2n^2}{8md^2}$$

The reverse of this process is then as follows:

1. Multiply n by $8md^2$ to give $8md^2n$.

2. Divide $8md^2n$ by $3h^2$ to give $\dfrac{8md^2n}{3h^2}$.

3. Take the square root of $\dfrac{8md^2n}{3h^2}$ to give

$$arcE(n) = \sqrt{\frac{8md^2n}{3h^2}}$$

7.4.4   Differentiation of fractional indices

When differentiation was introduced in Chapter 3, Section 3.5.1, on page 66, we saw that

$$\frac{d}{dx}(ax^n) = anx^{n-1}$$

In other words, if we differentiate an expression containing a power of x, we multiply by the power and reduce the power by 1. Although the examples of this given so far have all involved integral powers of x, this by no means has to be the case. So, for example,

$$\frac{d}{dx}\left(x^{\frac{2}{3}}\right) = \frac{2}{3}x^{\frac{2}{3}-1}$$

$$= \frac{2}{3}x^{-\frac{1}{3}}$$

This could also be written as

$$\frac{2}{3x^{\frac{1}{3}}}$$

because

$$x^{-n} = \frac{1}{x^n}$$

A second example is

$$\frac{d}{dx}(3x^{-0.25}) = 3 \times (-0.25)x^{-0.25-1}$$

$$= -0.75x^{-1.25}$$

Worked example 7.8

The density of states is the number of quantum states per unit energy at a given energy and is an important characteristic of a quantum system. For a system whose energy E is given in terms of a single quantum number n, the density of states $\rho(E)$ is given by

$$\rho(E) = \frac{dn}{dE}$$

Determine $\rho(E)$ for the particle in a one-dimensional box.

Chemical background

The density of states function gives us a way of measuring energy level spacing. For some systems this spacing will be constant, whereas for others it will vary. A further complication occurs in degenerate systems, where the degeneracy has to be included in the function as well as the spacing.

Solution to worked example 7.8

In order to calculate dn/dE we need to rearrange our expression for E to make n the subject. Starting from

$$E = \frac{n^2 h^2}{8ma^2}$$

we multiply both sides by $8ma^2$ to give

$$8ma^2E = n^2h^2$$

Now divide both sides by h^2 to give

$$\frac{8ma^2E}{h^2} = n^2$$

Finally, we can take the square root of each side to give

$$\frac{a}{h}(8mE)^{\frac{1}{2}} = n$$

We can rewrite this in the opposite order with the variable E separate as

$$n = \frac{a}{h}(8m)^{\frac{1}{2}} E^{\frac{1}{2}}$$

Before differentiating this, it may be useful to realize that

$$\frac{d}{dx}(x^{\frac{1}{2}}) = \frac{1}{2}x^{\frac{1}{2}-1} = \frac{1}{2}x^{-\frac{1}{2}}$$

so that

$$\frac{dn}{dE} = \frac{a}{h}(8m)^{\frac{1}{2}}\left(\frac{1}{2}E^{-\frac{1}{2}}\right)$$

Because

$$8^{\frac{1}{2}} = (4 \times 2)^{\frac{1}{2}} = 4^{\frac{1}{2}} \times 2^{\frac{1}{2}} = 2 \times 2^{\frac{1}{2}}$$

and

$$E^{-\frac{1}{2}} = \frac{1}{E^{\frac{1}{2}}}$$

we can write the expression for the density of states as

$$\frac{dn}{dE} = \frac{a}{h}(2 \times 2^{\frac{1}{2}})m^{\frac{1}{2}} \times \frac{1}{2E^{\frac{1}{2}}}$$

so that

$$\rho(E) = \left(\frac{a}{h}\right)\left(\frac{2m}{E}\right)^{\frac{1}{2}}$$

7.4.5   Use of standard integrals

In general the process of integration is more difficult than that of differentiation. As chemists, it therefore makes sense when faced with an integral to evaluate, to see whether the result is available, rather than starting from scratch. This is the idea behind many books that consist entirely of tabulations of such standard integrals. The trick in using these is to recognize which form is appropriate; frequently, this requires a certain degree of manipulation of the function to be integrated.

Suppose that we wish to determine the integral

$$\int_{0.0}^{0.5} \frac{dx}{\sqrt{1 - 2x^2}}$$

given the standard integral

$$\int \frac{dx}{\sqrt{a^2 - x^2}} = \sin^{-1}\left(\frac{x}{a}\right) + C$$

The first step is to recast the function to be integrated into the required form. This can be done by realizing that $2x^2$ needs to be become x^2, which can be realized by dividing the square root by 2 and balancing this by also dividing the top of the fraction by $\sqrt{2}$. This gives

$$\int_{0.0}^{0.5} \frac{\frac{1}{\sqrt{2}}\,dx}{\sqrt{\frac{1 - 2x^2}{2}}} = \frac{1}{\sqrt{2}} \int_{0.0}^{0.5} \frac{dx}{\sqrt{\frac{1}{2} - x^2}}$$

because the constant term $\dfrac{1}{\sqrt{2}}$ can be taken outside the integration. Comparing the function to be integrated with the general form now shows that $a^2 = \dfrac{1}{2}$ so that $a = \sqrt{\dfrac{1}{2}} = \dfrac{1}{\sqrt{2}}$. Applying the result for the standard integral shows that

$$\int_{0.0}^{0.5} \frac{dx}{\sqrt{1-2x^2}} = \sqrt{\frac{1}{2}} \left[\sin^{-1} \left(\frac{x}{\frac{1}{\sqrt{2}}} \right) \right]_{0.0}^{0.5}$$

$$= \sqrt{\frac{1}{2}} \left[\sin^{-1}(x\sqrt{2}) \right]_{0.0}^{0.5}$$

$$= \sqrt{\frac{1}{2}} \left[\sin^{-1} 0.5\sqrt{2} - \sin^{-1} 0.0\sqrt{2} \right]$$

$$= 0.7071 \left[\sin^{-1} 0.7071 - \sin^{-1} 0.0 \right]$$

$$= 0.7071 \left[0.7854 - 0.0000 \right]$$

$$= 0.5554$$

Worked example 7.9

The expectation value of the position of the particle within the box, $\langle x \rangle$, is given by the equation

$$\langle x \rangle = \frac{2}{a} \int_{0}^{a} x \sin^2 \left(\frac{n\pi x}{a} \right) dx$$

Determine the value of $\langle x \rangle$ given the standard integral

$$\int x \sin^2 ax\, dx = \frac{x^2}{4} - \frac{1}{4} \sin 2ax - \frac{1}{8} \cos 2ax + C$$

Chemical background

The expectation value of a quantity is its average experimental value. It clearly gives us a link between the sometimes abstract nature of quantum mechanics and the real world. In some cases, expectation values reinforce our classical view of systems, whereas in others they provide surprising results that involve us in further thought.

Solution to worked example 7.9

We begin by recasting the standard integral given into the form suggested by the particular problem. In this case the variable x is multiplied by the constant $\frac{n\pi}{a}$; we thus replace a in the standard integral so that it becomes

$$\int x \sin^2\left(\frac{n\pi x}{a}\right) dx = \frac{x^2}{4} - \frac{x}{4}\sin\left(\frac{2n\pi x}{a}\right) - \frac{1}{8}\cos\left(\frac{2n\pi x}{a}\right) + C$$

The definite integral we require is then

$$\int_0^a x \sin^2\left(\frac{n\pi x}{a}\right) dx = \left[\frac{x^2}{4} - \frac{x}{4}\sin\left(\frac{2n\pi x}{a}\right) - \frac{1}{8}\cos\left(\frac{2n\pi x}{a}\right)\right]_0^a$$

$$= \left[\frac{a^2}{4} - \frac{a}{4}\sin 2n\pi - \frac{1}{8}\cos 2n\pi\right] - \left[0 - 0 - \frac{1}{8}\cos 0\right]$$

Because n can only take integral values, $\sin 2n\pi = 0$, and $\cos 2n\pi = 1$. Therefore,

$$\int_0^a x \sin^2\left(\frac{n\pi x}{a}\right) dx = \left[\frac{a^2}{4} - 0 - \frac{1}{8}\times 1\right] - \left[-\frac{1}{8}\times 1\right]$$

$$= \frac{a^2}{4} - \frac{1}{8} + \frac{1}{8}$$

$$= \frac{a^2}{4}$$

Consequently,

$$\langle x \rangle = \frac{2}{a}\times\frac{a^2}{4}$$

$$= \frac{a}{2}$$

7.5   The free particle

Although a free particle that is not subject to any forces is not a particularly interesting system from a chemical point of view, it does allow us to work through a quantum mechanical treatment at a relatively simple level. A knowledge of the potential energy of the system allows the Schrödinger equation to be written and subsequently solved to give the general wavefunction. We then need to apply the boundary conditions and normalize the function before it can be used to predict properties.

7.5.1   The complex conjugate

We saw in Chapter 6, Section 6.5.1, on page 213 that a complex number z was defined in terms of real numbers a and b as

$$z = a + ib$$

where $i^2 = -1$. The complex conjugate of z, known as z^*, is defined as

$$z^* = a - ib$$

So, to obtain the conjugate of a complex number, we change the sign of the imaginary part. If $z = 2 + 3i$, we would then have $z^* = 2 - 3i$. Similarly, if $z = 2 - 4i$, we would have $z^* = 2 + 4i$. Therefore, z and z^* are called a *complex conjugate pair;* each one is the complex conjugate of the other.

   Notice that because

$$e^{ikx} = \cos kx + i \sin kx$$

and

$$e^{-ikx} = \cos kx - i \sin kx$$

e^{ikx} and e^{-ikx} are a complex conjugate pair.

7.5.2   The modulus of a complex number

We saw in the previous chapter that the modulus of a vector was its length, without regard to direction. Complex numbers can be represented in a similar fashion to vectors by use of an Argand diagram. This plots the real part a of a complex number in the x-direction, and the imaginary part b in the y-direction. Such a plot for the complex number $1 + 4i$ is shown in Figure 7.9.

   The modulus of the complex number z is then given as

$$|z| = \sqrt{a^2 + b^2}$$

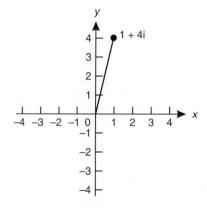

Figure 7.9   The Argand diagram for $1 + 4i$.

This can be compared with the expression for calculating the modulus of a vector which we met in Chapter 6, Section 6.4.3. If we now consider the product of z^*, the complex conjugate of z, with z, we get

$$z^*z = (a - ib)(a + ib)$$

$$= a(a + ib) - ib(a + ib)$$

$$= a^2 + aib - iba - i^2b^2$$

Because $aib = iba$, and $i^2 = -1$, we obtain

$$z^*z = a^2 + b^2$$

and taking the square root of both sides of this equation gives

$$\sqrt{z^*z} = \sqrt{a^2 + b^2} = |z|$$

so that the modulus of a complex number can be obtained by taking the square root of the product of the complex number and its complex conjugate.

Worked example 7.10

One wavefunction Ψ that satisfies the Schrödinger equation for a free particle that moves only in the x-direction is given by the equation

$$\Psi = A \exp\left(-i\sqrt{8\pi^2 mE}\,\frac{x}{h}\right)$$

where A is a constant, h is Planck's constant, m is the mass of the particle, and E its energy when at position x. The probability of finding the particle at a particular position is given by the function $\Psi^*\Psi$. Determine the equation that describes this function.

Chemical background

An example of a free particle in chemistry is an ionized electron, which may have any energy value. The Balmer series in the spectrum of hydrogen converges when such electrons fall to the energy level with principal quantum number 2.

Because the final value for the probability does not depend on x, the free particle has an equal probability of being found anywhere in the x-direction, and it is said to be nonlocalized.

Solution to worked example 7.10

From the expression for the wavefunction

$$\Psi = A \exp\left(-i\sqrt{8\pi^2 mE}\,\frac{x}{h}\right)$$

we can immediately write down an expression for the conjugate wavefunction

$$\Psi^* = A^* \exp\left(i\sqrt{8\pi^2 mE}\, \frac{x}{h} \right)$$

where A^* is the complex conjugate of A, and $-i$ has been replaced by i. We are now able to form the product of these two functions:

$$\Psi^*\Psi = A^* \exp\left(i\sqrt{8\pi^2 mE}\, \frac{x}{h} \right) A \exp\left(-i\sqrt{8\pi^2 mE}\, \frac{x}{h} \right)$$

and because

$$e^{-x} = \frac{1}{e^x}$$

this can be rewritten as

$$\Psi^*\Psi = A^*A \frac{\exp\left(i\sqrt{8\pi^2 mE}\, \dfrac{x}{h} \right)}{\exp\left(i\sqrt{8\pi^2 mE}\, \dfrac{x}{h} \right)}$$

$$= A^*A$$

$$= |A|^2$$

7.6   The hydrogen atom wavefunction

Wavefunctions can be defined for the electron in a hydrogen atom in both its ground state and in higher energy states. Their form is found by solving the appropriate Schrödinger equation, and details of this procedure can be found in physical chemistry textbooks. The wavefunctions are normally expressed as functions of three variables, which are the distance r of the electron from the nucleus, the zenith angle θ, and the azimuthal angle ϕ, as shown in Figure 7.10. These quantities are known as spherical polar coordinates and define the position of a particle in space, and are more useful than the more familiar Cartesian coordinates x, y, and z when we are dealing with systems such as this where the symmetry is spherical.

In its ground state, the single electron in a hydrogen atom occupies the 1s shell; this could be promoted to the 2s or 2p level in the excited state. The numbers 1 and 2 are known as the principal quantum number.

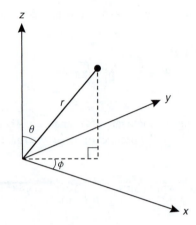

Figure 7.10   The definition of the spherical polar coordinates r, θ, and ϕ.

7.6.1   *Differentiation of a product*

We saw in Chapter 3, Section 3.5.1, on page 66, how to differentiate simple functions. We also come across *products* of functions that need to be differentiated, such as $x^2 e^x$. In this case, we need to be able to apply the standard rule for the differentiation of a product. If our function $f(x)$ can be expressed as the product of two functions $u(x)$ and $v(x)$, such that

$$f(x) = u(x)v(x)$$

then

$$\frac{df(x)}{dx} = v(x)\frac{du(x)}{dx} + u(x)\frac{dv(x)}{dx}$$

This is perhaps more easily remembered in words as "differentiate the first and multiply by the second, differentiate the second and multiply by the first."
In our preceding example, we then have

$$u(x) = x^2, \; v(x) = e^x$$

and

$$\frac{du(x)}{dx} = 2x \qquad \frac{dv(x)}{dx} = e^x \qquad\qquad \text{(Section 7.6.3)}$$

so

$$\frac{d(x^2 e^x)}{dx} = e^x \times 2x + x^2 \times e^x$$

$$= 2xe^x + x^2 e^x$$

7.6.2   Integration by parts

The equation expressing the rule for differentiating a product can also be of use to us when integrating certain more complicated functions. Writing this with the omission of the function notation for clarity as

$$\frac{d(uv)}{dx} = v\frac{du}{dx} + u\frac{dv}{dx}$$

can be rearranged to

$$u\frac{dv}{dx} = \frac{d(uv)}{dx} - v\frac{du}{dx}$$

by subtracting the term $v\dfrac{du}{dx}$ from both sides.

We can now integrate this expression:

$$\int u\frac{dv}{dx}dx = \int \frac{d(uv)}{dx}dx - \int v\frac{du}{dx}dx$$

Because integration is the reverse of differentiation, integrating a derivative gives back the original function, and so we have

$$\int \frac{d(uv)}{dx}dx = uv$$

which gives us

$$\int u\frac{dv}{dx}dx = uv - \int v\frac{du}{dx}dx$$

Here, u is a function that we will need to differentiate, and dv/dx is a function we will need to integrate. This equation is used to perform the integration process known as *integration by parts*. Now, consider the calculation of the integral

$$\int x \ln x dx$$

We can integrate and differentiate the term x, but with $\ln x$ we would prefer to differentiate. So, if we make the substitutions

$$u = \ln x \quad \text{and} \quad \frac{dv}{dx} = x$$

this gives, on differentiating u and integrating dv/dx,

$$\frac{du}{dx} = \frac{1}{x}$$

and
$$v = \frac{x^2}{2}$$

ignoring, for the time being, the constant of integration. We can now substitute into the general equation

$$\int u \frac{dv}{dx} dx = uv - \int v \frac{du}{dx} dx$$

to give

$$\int x \ln x \, dx = \frac{x^2}{2} \ln x - \int \frac{x^2}{2} \frac{1}{x} dx$$

$$= \frac{x^2}{2} \ln x - \int \frac{x}{2} dx$$

$$= \frac{x^2}{2} \ln x - \frac{x^2}{4} + C$$

where C is the constant of integration, normally inserted at the end of the calculation for simplicity. Be aware that it is easy to forget it!

7.6.3   Calculus of the exponential function

The exponential function e^x is the only function that does not change upon differentiation and integration. We now need to use the slightly more general result for the function e^{ax} where a is a constant. Examples are e^{3x} or e^{-2x}. The rules we need are

$$\frac{d(e^{ax})}{dx} = ae^{ax} \quad \text{and} \quad \int e^{ax} dx = \frac{e^{ax}}{a} + C$$

Some examples of these are

$$\frac{d(e^{-x})}{dx} = -e^{-x}$$

$$\frac{d(e^{4x})}{dx} = 4e^{4x}$$

$$\int e^{-3x} dx = \frac{e^{-3x}}{-3} + C$$

Worked example 7.11

During the normalization of the wavefunction representing the lowest electronic state of the hydrogen atom, the integral

$$\int_0^\infty r^2 \exp\left(-\frac{2r}{a_0}\right) dr$$

needs to be evaluated. Use integration by parts to evaluate this integral.

Chemical background

The process of normalization is used to ensure that the probability of finding the specified particle within a given volume is correct. Because the electron must be found somewhere within the hydrogen atom, the probability of finding it is 1, and so the probability function derived from the wavefunction must integrate over all space to give this value. It can be adjusted by the inclusion of a normalization factor, which is usually given the symbol N.

In some cases, the normalization constant can be represented by a generalized formula. This is the case for the harmonic oscillator, although the formula is relatively complicated.

Solution to worked example 7.11

The function to be integrated is shown in Figure 7.11. We are obviously going to use the preceding equation for performing integration by parts. This problem is given in terms of the variable r, so the equation we need to use is actually

$$\int u \frac{dv}{dr} dr = uv - \int v \frac{du}{dr} dr$$

when expressed in terms of this variable. We now need to assign the two functions $u(r)$ and $v(r)$ to the two functions r^2 and $\exp\left(-\frac{2r}{a_0}\right)$. If we set $u(r) = r^2$, then we obtain, by differentiation,

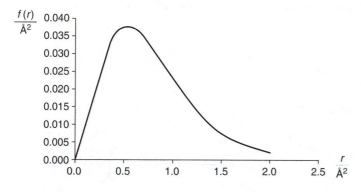

Figure 7.11   Graph of the function $f(r)$ integrated in worked example 7.11.

$$\frac{du}{dr} = 2r$$

whereas if we set $\dfrac{dv}{dr} = r^2$, we obtain, by integration,

$$v(r) = \frac{1}{3}r^3$$

The first of these seems to be leading towards a simpler calculation, but we also need to consider what happens to the second function.

If we have $u(r) = \exp\left(-\dfrac{2r}{a_0}\right)$, then, by differentiating, we obtain

$$\frac{du}{dr} = -\frac{2}{a_0}\exp\left(-\frac{2r}{a_0}\right)$$

whereas the substitution $\dfrac{dv}{dr} = \exp\left(-\dfrac{2r}{a_0}\right)$ gives, on integration,

$$v(r) = \frac{\exp\left(-\dfrac{2r}{a_0}\right)}{-\dfrac{2}{a_0}}$$

$$= -\frac{a_0}{2}\exp\left(-\frac{2r}{a_0}\right)$$

There is no particular advantage in either of these substitutions for the exponential term, so we choose arbitrarily to take

$$u(r) = r^2 \quad \text{and} \quad \frac{dv}{dr} = \exp\left(-\frac{2r}{a_0}\right)$$

so that

$$\frac{du}{dr} = 2r \quad \text{and} \quad v = -\frac{a_0}{2}\exp\left(-\frac{2r}{a_0}\right)$$

Making these substitutions into our general equation

$$\int u \frac{dv}{dr} dr = uv - \int v \frac{du}{dr} dr$$

gives

$$\int r^2 \exp\left(-\frac{2r}{a_0}\right) = r^2\left(-\frac{a_0}{2}\exp\left(-\frac{2r}{a_0}\right)\right) - \int\left(-\frac{a_0}{2}\exp\left(-\frac{2r}{a_0}\right)\right)2r\,dr$$

$$= -\frac{a_0}{2}r^2\exp\left(-\frac{2r}{a_0}\right) + a_0\int r\exp\left(-\frac{2r}{a_0}\right)dr$$

At this point in a problem involving the use of integration by parts, we would now evaluate the integral in the second term on the right-hand side of the equation, which is

$$\int r\exp\left(-\frac{2r}{a_0}\right)dr$$

In this case, however, it is not at all obvious how this can be done. A closer inspection reveals that this integral can also be evaluated using integration by parts. If we set

$$u = r \quad \text{and} \quad \frac{dv}{dr} = \exp\left(-\frac{2r}{a_0}\right)$$

we have

$$\frac{du}{dr} = 1 \quad \text{and} \quad v(r) = \frac{\exp\left(-\frac{2r}{a_0}\right)}{-\frac{2}{a_0}} = -\frac{a_0}{2}\exp\left(-\frac{2r}{a_0}\right)$$

Substituting these values into

$$\int u \frac{dv}{dr} dr = uv - \int v \frac{du}{dr} dr$$

gives us

$$\int r \exp\left(-\frac{2r}{a_0}\right) dr = -\frac{a_0}{2} r \exp\left(-\frac{2r}{a_0}\right) - \int\left(-\frac{a_0}{2}\exp\left(-\frac{2r}{a_0}\right)\right) \times 1\, dr$$

$$= -\frac{a_0}{2} r \exp\left(-\frac{2r}{a_0}\right) + \frac{a_0}{2}\int \exp\left(-\frac{2r}{a_0}\right) dr$$

$$= -\frac{a_0}{2} r \exp\left(-\frac{2r}{a_0}\right) + \frac{a_0}{2}\,\frac{\exp\left(-\dfrac{2r}{a_0}\right)}{-\dfrac{2}{a_0}}$$

$$= -\frac{a_0}{2} r \exp\left(-\frac{2r}{a_0}\right) - \frac{a_0^2}{4}\exp\left(-\frac{2r}{a_0}\right)$$

Because we have been able to evaluate this integral completely, the expression can now be substituted into our equation for the original integral to give

$$\int r^2 \exp\left(-\frac{2r}{a_0}\right) dr = -\frac{a_0}{2} r^2 \exp\left(-\frac{2r}{a_0}\right) + a_0\left(-\frac{a_0}{2} r \exp\left(-\frac{2r}{a_0}\right) - \frac{a_0^2}{4}\exp\left(-\frac{2r}{a_0}\right)\right)$$

$$= -\frac{a_0}{2} r^2 \exp\left(-\frac{2r}{a_0}\right) - \frac{a_0^2}{2} r \exp\left(-\frac{2r}{a_0}\right) - \frac{a_0^3}{4}\exp\left(-\frac{2r}{a_0}\right)$$

$$= \exp\left(-\frac{2r}{a_0}\right)\left(-\frac{a_0 r^2}{2} - \frac{a_0^2 r}{2} - \frac{a_0^3}{4}\right)$$

We have not yet included the limits on our integration sign. It is actually easier to include them only at the end, so from our previous working we have

$$\int_0^\infty r^2 \exp\left(-\frac{2r}{a_0}\right) dr = \left[\exp\left(-\frac{2r}{a_0}\right)\left(-\frac{a_0 r^2}{2} - \frac{a_0^2 r}{2} - \frac{a_0^3}{4}\right)\right]_0^\infty$$

Putting in an upper limit of $r = \infty$ first, and realizing that

$$\exp\left(-\frac{2r}{a_0}\right) = \frac{1}{\exp\left(\dfrac{2r}{a_0}\right)}$$

because $e^{-x} = 1/x$, shows that the whole of this term will tend to zero for the upper limit because we have the reciprocal of a very large number.

Multiplying by zero gives a value of zero for the whole expression as $r \rightarrow \infty$.

On the other hand, when $r = 0$,

$$\exp\left(-\frac{2r}{a_0}\right) = e^0 = 1$$

because any number raised to the power zero is one. The first two terms in brackets are zero when $r = 0$, and so we are left with

$$\int_0^\infty r^2 \exp\left(-\frac{2r}{a_0}\right) dr = 0 - \left(-\frac{a_0^3}{4}\right)$$

$$= \frac{a_0^3}{4}$$

7.6.4   Multiple integration

In Chapter 3, Section 3.7.5, on page 92, we met the idea of a partial derivative, where a function $f(x,y)$ of two variables can be differentiated with respect to either variable while keeping the other constant. Differentiating $f(x,y)$ with respect to x while keeping y constant gives the derivative denoted by

$$\left(\frac{\partial f}{\partial x}\right)_y$$

whereas differentiating $f(x,y)$ with respect to y while keeping x constant leads to the derivative noted by

$$\left(\frac{\partial f}{\partial y}\right)_x$$

When we consider integration of this function, we can determine

$$\int_a^b f(x, y)dx$$

where $f(x,y)$ is integrated with respect to x while keeping y constant, or

$$\int_a^b f(x, y)dy$$

where $f(x,y)$ is integrated with respect to y while keeping x constant. In both these cases the lower and upper integration limits are assumed to be a and b, respectively.

Alternatively, we can also determine the double integral of $f(x,y)$ with respect to both x and y. This would be denoted as

$$\int_{c}^{d}\int_{a}^{b} f(x,y)dxdy$$

By convention this is read from the outside in. Thus the constants c and d are the lower and upper limits, respectively, when integrating with respect to y, whereas a and b are the respective lower and upper limits when integrating with respect to x. The process involved can be made clearer by rewriting the expression using brackets to give

$$\int_{c}^{d}\left(\int_{a}^{b} f(x,y)dx\right)dy$$

Using the convention that we evaluate quantities in brackets first, we begin by integrating $f(x,y)$ with respect to x between limits a and b, and then integrating the result of this with respect to y between limits c and d.

Applying this to a double integral such as

$$\int_{0}^{1}\int_{2}^{3} (3xy + 2x)dxdy$$

we begin by integrating $3xy + 2x$ with respect to x to give

$$\frac{3x^2 y}{2} + \frac{2x^2}{2} = \frac{3x^2 y}{2} + x^2$$

treating y as a constant. Substituting this into the original expression, and using the standard notation for integration, gives

$$\int_{0}^{1}\int_{2}^{3} (3xy + 2x)dxdy = \int_{0}^{1}\left[\frac{3x^2 y}{2} + x^2\right]_{2}^{3} dy$$

Substituting the integration limits gives

$$\int_{0}^{1}\left[\frac{3x^2 y}{2} + x^2\right]_{2}^{3} dy = \int_{0}^{1}\left\{\left[\frac{3\times 3^2 y}{2} + 3^2\right] - \left[\frac{3\times 2^2 y}{2} + 2^2\right]\right\}dy$$

$$= \int_{0}^{1}\left\{\left(\frac{9y}{2} + 9\right) - (6y + 4)\right\}dy$$

$$= \int_{0}^{1}\left(5 - \frac{3y}{2}\right)dy$$

This can now be integrated with respect to y in the usual fashion to give

$$\int_0^1 \left(5 - \frac{3y}{2}\right)dy = \left[5y - \frac{3y^2}{4}\right]_0^1$$

$$= \left[5\times1 - \frac{3\times1}{4}\right] - \left[0\right]$$

$$= 5 - \frac{3}{4}$$

$$= \frac{17}{4}$$

Consequently, the overall result is that

$$\int_0^1 \int_2^3 (3xy + 2x)dxdy = \frac{17}{4}$$

It is also possible to evaluate triple integrals in a similar fashion.

7.6.5 Calculus of the trigonometric functions

The trigonometric functions and their inverses were introduced in Chapter 6, Section 6.2, on page 194. In particular, Figure 6.5 showed the graphs of the sine and cosine functions. Because we can plot these graphs, it follows that we can determine the gradient of the function at any point. This, as we saw in Chapter 3, Section 3.5.1, on page 66, is equivalent to determining the derivative. Consequently, we can discuss the calculus of the trigonometric functions.

In this section, we will merely note the results of differentiating and integrating the general sine and cosine functions:

$$\frac{d}{dx}(\sin(ax+b)) = a\cos(ax+b)$$

$$\frac{d}{dx}(\cos(ax+b)) = -a\sin(ax+b)$$

$$\int \sin(ax+b)dx = -\frac{1}{a}\cos(ax+b) + C$$

$$\int \cos(ax+b) = \frac{1}{a}\sin(ax+b) + C$$

where C is the constant of integration.

As an example of differentiation, consider

$$\frac{d}{dx}(\cos(3x - 2))$$

Comparing with the general form of the function $\cos(ax + b)$ we see that we have

$$a = 3$$

$$b = -2$$

and so

$$\frac{d}{dx}(\cos(3x - 2)) = -3\sin(3x - 2)$$

noting in particular that the term on the right is negative. Now consider

$$\int_0^\pi \cos 5x dx$$

Comparing with the general function $\cos(ax + b)$ shows that

$$a = 5$$

$$b = 0$$

and so

$$\int \cos 5x dx = \frac{1}{5}\sin 5x$$

However, we have been asked to determine the definite integral, so we need to consider the lower and upper limits, which are 0 and π, respectively. Unless specified, such limits on trigonometric functions are taken as being in radians. We then have

$$\int_0^\pi \cos 5x dx = \left[\frac{1}{5}\sin 5x\right]_0^\pi$$

$$= \frac{1}{5}\sin 5\pi - \frac{1}{5}\sin 0$$

$$= 0 - 0$$

$$= 0$$

because the value of the sine of any multiple of π rad (180°) is zero.

Worked example 7.12

Determine the integral

$$\int_0^{2\pi}\int_0^{\pi}\int_0^{\infty} e^{-\frac{2r}{a_0}} r^2 \sin\theta\, dr\, d\theta\, d\phi$$

which is required in order to evaluate the normalization constant for the hydrogen atom wavefunction.

Chemical background

Notice that this equation is formulated in terms of the spherical polar coordinates introduced at the start of the section. As we have seen earlier in this section, the normalization constant is introduced in order to ensure that the probability of finding an electron anywhere in space is exactly 1. This is expressed mathematically by the equation

$$\int_{-\infty}^{\infty} \Psi^*\Psi = 1$$

where Ψ is the wavefunction and Ψ^* its complex conjugate, the integration limits being taken from minus infinity to infinity to represent the whole of space.

Solution to worked example 7.12

We can begin by rewriting the expression to be integrated by changing the order of the terms and using brackets, as

$$\int_0^{2\pi}\left(\int_0^{\pi} \sin\theta \left(\int_0^{\infty} r^2 e^{-\frac{2r}{a_0}} dr\right) d\theta\right) d\phi$$

Working from the outside in, the integral of $d\phi$ is ϕ, so we have

$$\int_0^{2\pi}\left(\int_0^{\pi} \sin\theta \left(\int_0^{\infty} r^2 e^{-\frac{2r}{a_0}} dr\right) d\theta\right) d\phi = \left(\int_0^{\pi} \sin\theta \left(\int_0^{\infty} r^2 e^{-\frac{2r}{a_0}} dr\right) d\theta\right)\left[\phi\right]_0^{2\pi}$$

$$= 2\pi\left(\int_0^{\pi} \sin\theta \left(\int_0^{\infty} r^2 e^{-\frac{2r}{a_0}} dr\right) d\theta\right)$$

once the integration limits for ϕ are inserted.

As we saw earlier,

$$\int \sin x dx = -\cos x + C$$

so that

$$\int_0^\pi \sin \theta d\theta = \left[-\cos \theta \right]_0^\pi$$

$$= \left[-\cos \pi \right] - \left[-\cos 0 \right]$$

$$= [-(-1)] - [-1]$$

$$= 1 + 1$$

$$= 2$$

remembering that θ is, in this case, expressed in radians, and that the two adjacent negative signs combine to give a positive. Our integral is now

$$2\pi \times 2 \left(\int_0^\infty r^2 e^{-\frac{2r}{a_0}} dr \right)$$

The result of worked example 7.11 now allows us to write

$$\int_0^{2\pi} \left(\int_0^\pi \sin \theta \left(\int_0^\infty r^2 e^{-\frac{2r}{a_0}} dr \right) d\theta \right) d\phi = 2\pi \times 2 \times \frac{a_0^3}{4}$$

$$= \pi a_0^3$$

7.7   The helium atom

The quantum mechanics of the helium atom are rather more complicated than those of the hydrogen atom because we now have to account for the interactions between electrons as well as that between an electron and the nucleus. These interactions are illustrated in Figure 7.12. In practice, we build a trial wavefunction based on the information we have already obtained for the 1s orbital of hydrogen.

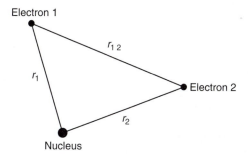

Figure 7.12   Interactions between particles in the helium atom.

7.7.1   Stationary points

The method for determining the stationary points of a function has been presented in Chapter 4, Section 4.6.2, on page 127. The stationary points of a function $f(x)$ occur when

$$\frac{df(x)}{dx} = 0$$

and

$$\frac{d^2 f(x)}{dx^2} < 0 \text{ for a maximum}$$

while

$$\frac{d^2 f(x)}{dx^2} > 0 \text{ for a minimum.}$$

Worked example 7.13

The energy E of the helium atom can be expressed as

$$E = \left(\frac{-2\pi^2 m_e e^4}{h^2} \right) \left(-2(Z')^2 + \frac{27}{4} Z' \right)$$

where m_e is the mass of an electron, e is the electronic charge, and h is Planck's constant. Z' is the effective nuclear charge on the helium atom and is treated as an adjustable parameter. Find the value of Z' that makes the value of E a minimum.

Chemical background

This approach leads to a value for the energy of −77.5 eV, which compares with an experimental value of −79.0 eV. A calculated value of

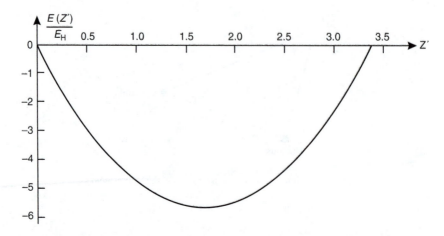

Figure 7.13   Graph of the function representing the energy of the helium atom. $E_H = 27.2$ eV.

The method being used here is actually an example of the variation theorem, which finds a number of applications in quantum mechanics. It involves specifying a trial wavefunction and then optimizing a specified adjustable parameter.

−78.7 eV can be obtained by including a term for electron correlation in the 1s orbital trial wavefunction. The function for E used here is shown in Figure 7.13.

Solution to worked example 7.13

We will obtain a minimum value of E when

$$\frac{dE}{dZ'} = 0 \quad \text{and} \quad \frac{d^2E}{dZ'^2} > 0$$

The term in the first bracket of the expression for E is a constant, and so this remains as a multiplier throughout the calculation. We know that

$$\frac{d(x^2)}{dx} = 2x \quad \text{and} \quad \frac{d(x)}{dx} = 1$$

so that

$$\frac{d}{dZ'}\left(-2(Z')^2 + \frac{27}{4}Z'\right) = -4Z' + \frac{27}{4}$$

This term will be zero when

$$-4Z' + \frac{27}{4} = 0$$

Adding $4Z'$ to both sides and dividing both sides by 4 isolates Z' to give

$$Z' = \frac{27}{16}$$

The full expression for the first derivative of E is

$$\frac{dE}{dZ'} = \left(\frac{-2\pi^2 m_e e^4}{h^2} \right) \left(-4Z' + \frac{27}{4} \right)$$

so the second derivative will be

$$\frac{d^2 E}{dZ'^2} = -4 \left(\frac{-2\pi^2 m_e e^4}{h^2} \right) = \frac{8\pi^2 m_e e^4}{h^2}$$

This is a positive value, and the second derivative being positive confirms that this stationary point is a minimum.

Exercises

1. Determine the results of the following computations, expressing your answers to the appropriate number of figures:

   (a)   9.453 287 712 – 0.446 76

   (b)   3.162 623 19 + 5.042 084 282 87

   (c)   0.622 844 4 × 12.309 421 722 306 987

   (d) $\dfrac{83.5919146149277511}{4.706508322}$

2. From the following data deduce the relationship between the variables x and y:

x	1.42	3.62	4.75	6.24	7.08
y	7.12	2.79	2.13	1.62	1.43

3. Two quantities p and q are directly proportional such that

   $$\frac{p}{q} = 42.7$$

   (a)   What is the value of p when $q = 14.3$?

   (b)   What is the value of q when $p = 0.314$?

4. Determine the stationary points of the function

$$f(x) = 2e^{3x} - 4x$$

5. Determine the location and the nature of the stationary points of the function $f(x) = 2x^2 - 4x + 8$. Hence state the location and nature of the stationary points of $g(x) = 2x^2 - 4x + 2$.

6. Determine the integral

$$\int_1^4 \int_{-1}^1 (x^2 y + 3xy) dx dy$$

7. Determine the integral

$$\int_1^2 \int_0^1 \int_2^3 \left(e^x + \frac{1}{z} - \frac{xy}{z^2} \right) dx dy dz$$

8. Give the conjugates of the following complex numbers:

   (a) $4 + 6i$

   (b) $-2 - i$

   (c) $9 - 4i$

9. Calculate the product $z_1 z_2$ if

$$z_1 = 3(4 - 3i)$$

   and

$$z_2 = 2(1 - 2i)$$

10. A sequence is defined by

$$f(n) = 4n^2 - 2n$$

   Give the first five terms of this sequence.

11. Find the next two terms of these sequences:

   (a)   4, 7, 10, 13, ...

   (b)   5, 10, 17, 26, ...

   (c) $-4, -3, 1, 10, ...$

12. Find the general term of the sequence

$$3, 6, 12, 24, 48, ...$$

13. Find the inverse of the function $f(x) = (2x+1)^2$.

14. Find the inverse of the function $f(x) = 2e^{ax^2}$.

15. Determine the following derivatives:

(a) $\dfrac{d}{dx}(x^{0.3})$

(b) $\dfrac{d}{dx}(2x^{-0.6})$

(c) $\dfrac{d}{dy}\left(\dfrac{-4}{y^{1.5}}\right)$

(d) $\dfrac{d}{dt}\left(\dfrac{1}{2t^{\frac{1}{2}}}\right)$

16. Differentiate these functions:

(a) $f(x) = x^2 \ln x$

(b) $g(x) = x\, e^{2x}$

(c) $h(y) = y^2\, e^{-4y}$

17. Differentiate the functions:

(a) e^{2x}

(b) e^{-3x}

(c) $\exp(3x+2)$.

18. Evaluate the following integrals:

(a) $\displaystyle \int e^{4x}\, dx$

(b) $\displaystyle \int_{0}^{3} e^{2x}\, dx$

(c) $\displaystyle \int_{-1}^{2} e^{-6x}\, dx$

19. Use the method of integration by parts to evaluate this definite integral:

$$\int_0^1 xe^{2x}\,dx$$

20. By writing $\ln x$ as the product of 1 and $\ln x$, evaluate the integral

$$\int \ln x\,dx$$

using the method of integration by parts.

Problems

1. The Bohr radius a_0 is defined as

$$a_0 = \frac{4\pi\varepsilon_0 \hbar^2}{m_e e^2}$$

where ε_0 is the permittivity of a vacuum, m_e is the electron rest mass, and e is the elementary charge. The constant $\hbar$ is related to the Planck constant h by the equation

$$\hbar = \frac{h}{2\pi}$$

Determine the value of a_0 to an appropriate number of figures, given the values

$$\varepsilon_0 = 8.854\ 187\ 816 \times 10^{-12}\ \text{F m}^{-1}$$

$$h = 6.626\ 075\ 5 \times 10^{-34}\ \text{J s}$$

$$m_e = 9.109\ 389\ 7 \times 10^{-31}\ \text{kg}$$

$$e = 1.602\ 177\ 33 \times 10^{-19}\ \text{C}$$

You will need to select an appropriate value of π that retains the level of precision implied by this data.

2. A function that can be used to describe the interaction energy $E(r)$ between a pair of molecules a distance r apart is

$$E(r) = \frac{A}{r^9} - \frac{B}{r^6}$$

where A and B are constants. Locate and identify the stationary points of this function.

3. The energy $E(n_x, n_y)$ of a particle in a two-dimensional box is given by the equation

$$E(n_x, n_y) = \frac{h^2}{8m}\left(\frac{n_x^2}{a^2} + \frac{n_y^2}{b^2}\right)$$

where n_x and n_y are quantum numbers in the x and y directions, respectively, a and b are the dimensions of the box in the corresponding directions, h is Planck's constant, and m is the mass of the particle. Calculate the inverse function arc $E(n)$ in the special case when $n_x = n_y = n$, and $a = b = d$.

4. The following data were obtained for the maximum kinetic energy T_{max} of the electrons emitted from a sodium surface in a photoelectric experiment as a function of frequency ν. Determine the relationship between T_{max} and ν without using a graph.

$\dfrac{T_{max}}{eV}$	0.0	0.5	1.0	1.5	2.0	2.5
$\dfrac{10^{-14}\nu}{Hz}$	5.47	6.73	7.99	9.25	10.51	11.77

5. The wavefunction $\Psi_{1s}(r)$ of the 1s orbital of the hydrogen atom is given by the equation

$$\Psi_{1s}(r) = 2\left(\frac{Z}{a_0}\right)^{\frac{3}{2}} \exp\left(-\frac{Zr}{a_0}\right)$$

where Z is the nuclear charge, a_0 the Bohr radius, and r the distance of the electron from the nucleus. Determine the inverse function arc $\Psi_{1s}(r)$.

6. The energy $E(m)$ for the particle on a ring is given in terms of the quantum number m as

$$E(m) = \frac{m^2 h^2}{8\pi^2 I}$$

where I is the moment of inertia and $m = 0, +\text{-}1, +\text{-}2 \dots$. Determine the density of states dm/dE for the particle on a ring.

7. The wavefunctions of the particle on a ring are of the form

$$\Psi_m = Ae^{im\phi}$$

where A is a normalization constant. Determine the value of A if the wavefunction satisfies the normalization condition

$$\int_0^{2\pi} \Psi_m^* \Psi_m d\phi = 1$$

where Ψ_m^* is the complex conjugate of Ψ_m.

8. The Variation Theorem is often used to determine the wavefunctions of the helium atom. However, it can also be used with a trial wavefunction for the one-dimensional particle in a box. This involves the integral

$$\int_0^a (ax - x^2)\left(-\frac{h^2}{8\pi^2 m}\right)\frac{d^2}{dx^2}(ax - x^2)$$

where a is the length of the box and x is the position of the particle of mass m within it. The constant h is Planck's constant. Perform the required differentiation of the function $ax - x^2$ and then evaluate this integral between the limits shown.

9. The expectation value of x^2 in the one-dimensional particle in a box is given by

$$\langle x^2 \rangle = \frac{2}{a}\int_0^a x^2 \sin^2\left(\frac{n\pi x}{a}\right)dx$$

Determine the value of $\langle x^2 \rangle$ given the standard integral

$$\int x^2 \sin^2 ax\, dx = \frac{x^3}{6} - \left(\frac{x^2}{4} - \frac{1}{8}\right)\sin 2ax - \frac{x\cos 2ax}{4} + C$$

10. The average distance of an electron from the nucleus in the 1s orbital of the hydrogen atom is given by the expression

$$\langle r_{1s} \rangle = \frac{1}{\pi a_0^3}\int_0^{2\pi}\int_0^{\infty}\int_0^{\pi} r^3 e^{-\frac{2r}{a_0}}\sin\theta\, d\theta\, dr\, d\phi$$

Determine the value of $\langle r_{1s} \rangle$ given the standard integral

$$\int_0^{\infty} x^3 e^{-kx}\, dx = \frac{6}{k^4}$$

Spectroscopy 8

The subject of spectroscopy provides us with a description of how chemical particles interact with electromagnetic radiation. Spectroscopic techniques can be classified according to the type of incident radiation (such as ultraviolet or infrared), or according to the molecular phenomena giving rise to the spectrum (such as rotation or vibration). Because we are concerned with interactions at the microscopic level, it is not surprising that the ideas of quantum mechanics are important in a development of this subject. In mathematical terms, most aspects of spectroscopy can be dealt with using techniques we have already met.

8.1   Calculation of dipole moments

For a heteronuclear diatomic molecule with excess charge q on each atom and an interatomic separation of r (Figure 8.1), the dipole moment μ is given by

$$\mu = qr$$

Figure 8.1   Charges of magnitude q separated by distance r forming a dipole.

Worked example 8.1

Calculate the dipole moment of CH, given that the excess charge on each atom is $0.29e$ and the bond length 1.124 Å.

Chemical background

For most second-row hydrides the hydrogen atom will denote the positive end of the dipole. The exceptions are LiH and BeH, and in these cases the dipole would be given a negative value.

Partial charges cannot be measured directly, but they can be determined by quantum mechanical calculations at various levels of sophistication. The resulting dipole moment is then an indication of their accuracy.

Dipole moments are usually given in units of debye (D). The debye is a non-SI unit that is equivalent to 3.336×10^{-30} C m.

Solution to worked example 8.1

Notice first of all that the excess charge q is given in terms of the electronic charge e, which has the value 1.602×10^{-19} C. Therefore,

$$q = 0.29 \times 1.602 \times 10^{-19} \text{ C} = 4.646 \times 10^{-20} \text{ C}$$

At this point we will carry one more figure than is strictly justified in order to avoid intermediate rounding errors.

The interatomic distance is given in terms of the non-SI unit Ångström, where $1 \text{ Å} = 10^{-10}$ m. We now have

$$\mu = qr$$

$$= 4.646 \times 10^{-20} \text{ C} \times 1.124 \times 10^{-10} \text{ m}$$

$$= 5.222 \times 10^{-30} \text{ C m}$$

To express this answer in D, we rearrange

$$1 \text{ D} = 3.33564 \times 10^{-30} \text{ C m}$$

by dividing both sides by 3.33564×10^{-30} so that

$$1 \text{C m} = \frac{1}{3.33564 \times 10^{-30}} \text{D}$$

$$= 2.998 \times 10^{29} \text{D}$$

Consequently,

$$\mu = 5.222 \times 10^{-30} \times 2.998 \times 10^{29} \text{ D}$$

$$= 1.566 \text{ D}$$

Returning to the original data, we see that the lowest number of significant figures is 2. Because all the arithmetic operations have been multiplication and division, the final value of μ should be given to 2 significant figures. The value to quote is therefore 1.6 D.

8.2   Dipole and quadrupole moments

The total electric potential $\phi(r)$ for the charge distribution a distance r and angle θ from the axis of a linear molecule is given by

$$\phi(r) = \frac{q}{4\pi\varepsilon_0 r} + \frac{p\cos\theta}{4\pi\varepsilon_0 r^2} + \frac{\Theta(3\cos^2\theta - 1)}{8\pi\varepsilon_0 r^3} + \dots$$

where q is the net charge, p the dipole moment, and Θ the quadrupole moment. The constant ε_0 is the permittivity of free space.

If the linear molecule consists of point charges, then

$$p = \Sigma_i q_i z_i \quad \text{and} \quad \Theta = \sum_i q_i z_i^2$$

where z_i represents distance along the molecular axis, measured from the centre of mass.

Worked example 8.2

Calculate the dipole and quadrupole moments for HF that has a bond length 0.917 Å and partial charges 0.44e.

Chemical background

The fact that HF is quite polar is reflected in the relatively high value of the partial charges.

Solution to worked example 8.2

It is straightforward to convert the partial charges to SI units. This gives

$$q = 0.44 \times 1.602 \times 10^{-19} \text{ C}$$

$$= 7.049 \times 10^{-20} \text{ C}$$

The appropriate values of z are less straightforward to determine. Because the relative atomic masses of hydrogen and fluorine are 1.008 and 18.998, respectively, we need to divide the interatomic distance 0.917 Å in this ratio. To do this we add the masses to give 20.006. The centre of mass will be closer to the heavier fluorine atom, so

$$z_F = \frac{1.008}{20.006} \times 0.917 \text{Å}$$

$$= 0.0462 \text{Å}$$

and

$$z_H = \frac{18.998}{20.006} \times 0.917 \text{Å}$$

$$= 0.8708 \text{Å}$$

The dipole moment is given by

$$p = \Sigma_i q_i z_i$$

$$= q_H z_H + q_F z_F$$

$$= 7.049 \times 10^{-20} \text{ C} \times 0.8708 \times 10^{-10} \text{ m}$$

$$+ (-7.049 \times 10^{-20} \text{ C}) \times (-0.0462 \times 10^{-10} \text{ m})$$

$$= 6.138 \times 10^{-30} \text{ C m} + 0.326 \times 10^{-30} \text{ C m}$$

$$= 6.464 \times 10^{-30} \text{ C m}$$

This can be converted to units of D, as in the previous example, to give

$$6.464 \times 10^{-30} \times 2.998 \times 10^{29} \text{ D} = 1.94 \text{ D}$$

Notice that, in calculating the dipole moment, we have applied the precedence rules as introduced in Chapter 1, Section 1.2, on page 5.

The quadrupole moment is given by

$$\Theta = \sum_i q_i z_i^2$$

$$= q_H z_H^2 + q_F z_F^2$$

$$= 7.049 \times 10^{-20} \text{ C} \times (0.8708 \times 10^{-10} \text{ m})^2$$

$$+ (-7.049 \times 10^{-20} \text{ C}) \times (0.0462 \times 10^{-10} \text{ m})^2$$

$$= 5.345 \times 10^{-40} \text{ C m}^2 - 0.01505 \times 10^{-40} \text{ C m}^2$$

$$= 5.330 \times 10^{-40} \text{ C m}^2$$

again using the precedence rules.

8.3   Electromagnetic radiation

Electromagnetic radiation is so called because it consists of an oscillating electric field and an oscillating magnetic field at right angles to one another. The electromagnetic spectrum ranges from gamma rays with a wavelength of about 10^{-12} m through to radio waves whose wavelength is approximately 100 m. In between, in ascending order of wavelength, are x-rays, ultraviolet radiation, visible light, infrared radiation, microwaves, and television waves (Figure 8.2). As well as being characterized by wavelength λ, electromagnetic radiation can be expressed in terms of its frequency ν.

8.3.1   Direct and inverse proportion

These types of proportion were discussed in Chapter 3, Section 3.7.3, on page 88.

Worked example 8.3

The relationship between the wavelength λ and frequency ν of electromagnetic radiation is

$$c = \lambda \nu$$

where c is its velocity, which can be taken as a constant. How can the relationship between λ and ν be described?

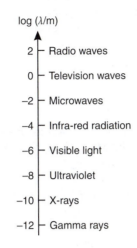

log (λ/m)

2 ⊢ Radio waves

0 ⊢ Television waves

−2 ⊢ Microwaves

−4 ⊢ Infra-red radiation

−6 ⊢ Visible light

−8 ⊢ Ultraviolet

−10 ⊢ X-rays

−12 ⊢ Gamma rays

Figure 8.2   The electromagnetic spectrum.

The velocity of light is exactly 299 792 458 m s^{-1}. As well as being measured experimentally, this value can be predicted from Maxwell's equations that describe the behaviour of electric and magnetic fields.

Chemical background

The energy E of a photon is given by the equation

$$E = h\nu$$

where h is Planck's constant. Therefore, as the frequency of the radiation increases, so does its energy. Because the wavelength is inversely proportional to the frequency, it also follows that higher wavelength radiation will have a lower energy.

An example is that of ultraviolet light, which has a lower energy and frequency than that of visible light. Consequently, it is more effective in inducing photochemical reactions.

Solution to worked example 8.3

Because the product of the two variables is a constant, we should be able to recognize instantly that they are inversely proportional. Alternatively, we can rearrange the equation given, by dividing both sides by λ, so that λ appears on one side and ν on the other.

$$\frac{c}{\lambda} = \nu$$

It should now be apparent that as we increase λ, then ν must decrease.

Worked example 8.4

In addition to wavelength and frequency, a quantity often used in spectroscopy is wavenumber. This is normally given the symbol $\tilde{\nu}$, defined as the reciprocal of the wavelength, i.e.,

$$\tilde{\nu} = \frac{1}{\lambda}$$

and has the units cm^{-1}. What is the wavenumber of the sodium D line having frequency 5.086×10^8 MHz? Take the velocity of electromagnetic radiation to be 2.998×10^8 m s^{-1}.

Chemical background

This type of discharge is typically seen in certain forms of street lighting.

When excited by an electric discharge, sodium vapour produces an emission spectrum that includes a yellow line at a wavelength of 589 nm. On closer analysis, this is seen to consist of two very closely spaced lines (a doublet) of wavelengths 589.76 nm and 589.16 nm, respectively.

Solution to worked example 8.4

We need to calculate the value of the quantity $1/\lambda$. If we start with the defining equation

$$c = \lambda \nu$$

we can divide both sides by ν to give

$$\frac{c}{\nu} = \lambda$$

As we wish to calculate $1/\lambda$, we now need to take the reciprocal of each side of this equation. The reciprocal of A is $1/\lambda$, and the reciprocal of any fraction is obtained by "turning it upside down." This now gives us

$$\frac{1}{\lambda} = \frac{\nu}{c}$$

$$= \frac{5.086 \times 10^8 \, \text{MHz}}{2.998 \times 10^8 \, \text{m s}^{-1}}$$

The first thing to notice is that both the top and the bottom of this fraction contain a 10^8 term, which can be cancelled. We then need to rewrite the unit MHz in a form that is of more use to us. The basic frequency unit Hz is equal to s^{-1}, and the prefix M means 10^6. This now gives

$$\frac{1}{\lambda} = \frac{5.086 \times 10^6 \, \text{s}^{-1}}{2.998 \, \text{m s}^{-1}}$$

The s^{-1} unit now appears in both the top and bottom of the right side of this equation, and can be cancelled. Using a calculator we now obtain

$$\frac{1}{\lambda} = 1.696 \times 10^6 \, m^{-1}$$

Although this is an acceptable unit for wavenumber, it is more usual to use units of cm^{-1}. We can do this if we realize that

$$100 \, cm = 1 \, m$$

or more usefully here,

$$10^2 \, cm = 1 \, m$$

We replace m in our expression with 10^2 cm. This gives

$$\frac{1}{\lambda} = 1.696 \times 10^6 \, m^{-1}$$

$$= 1.696 \times 10^6 (10^2 \, cm)^{-1}$$

The power of −1 is applied to every term in the bracket, so we need to use the fact that $(10^2)^{-1} = 10^{-2}$ to give

$$\frac{1}{\lambda} = 1.696 \times 10^6 \times 10^{-2} \, cm^{-1}$$

$$= 1.696 \times 10^4 \, cm^{-1}$$

8.4   The Beer–Lambert law

When radiation is incident upon some medium, it will be absorbed to a certain extent, as shown in Figure 8.3. This will result in its intensity being reduced,

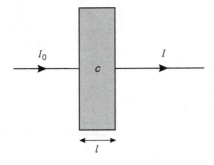

Figure 8.3   The absorption of radiation.

and such behaviour is described by the Beer–Lambert law, which can be expressed by the equation

$$A = \varepsilon c l$$

where ε is a constant known as the absorption coefficient, c is the concentration of the absorbing medium, and l is the distance through which the light travels within the medium. The quantity A is known as the absorbance, and is defined as

$$A = \log \frac{I_0}{I}$$

The distance l is commonly known as the path length.

where I_0 is the intensity of the incident radiation and I is the intensity once it has passed through the medium.

8.4.1   Rules of logarithms

These were discussed in Chapter 5, Section 5.4.2, on page 149.

Worked example 8.5

The fraction of incident radiation transmitted is the transmittance T, which is defined by the equation

$$T = \frac{I}{I_0}$$

Obtain an equation that relates the percentage transmittance T_{100} to the absorbance A.

Chemical background

Spectrophotometers are generally designed to work in one region of the spectrum only. Specific wavelengths are produced by using a rotating prism or a diffraction grating.

Many spectrophotometers are capable of displaying both absorbance and transmittance, and the relationship between these quantities is not immediately obvious. Generally, absorbance is the preferred quantity because of its proportionality to both concentration and path length.

Solution to worked example 8.5

The percentage transmittance T_{100} is related to the transmittance T (a fractional value) by the simple expression

$$T_{100} = 100T$$

$$= 100 \frac{I}{I_0}$$

This rearranges to give

$$\frac{I_0}{I} = \frac{100}{T_{100}}$$

which can be substituted into the expression for A:

$$A = \log\frac{I_0}{I}$$

$$\doteq \log\frac{100}{T_{100}}$$

Because we know that

$$\log\left(\frac{X}{Y}\right) = \log X - \log Y$$

we can rearrange the right-hand side of this equation to give

$$A = \log 100 - \log T_{100}$$

and because log 100 = 2, we obtain

$$A = 2 - \log T_{100}$$

which finally can be rearranged to give

$$\log T_{100} = 2 - A$$

8.5 Rotational spectroscopy

Rotating molecules give rise to spectra in the microwave region. It is possible to define three mutually perpendicular principal axes of rotation that pass through the centre of gravity of a molecule, and there are consequently three corresponding moments of inertia that depend on the molecular shape. The treatment of rotational spectra is consequently different for linear molecules (such as HCl), symmetric tops (such as CH_3F), spherical tops (such as CH_4), and asymmetric tops (such as H_2O). These different types of molecule are shown in Figure 8.4.

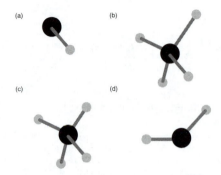

Figure 8.4   Molecules classified according to their rotational properties: (a) HCl (linear); (b) CH_3F (symmetric top); (c) CH_4 (spherical top); and (d) H_2O (asymmetric top).

8.5.1   Sequences

This subject was discussed in Chapter 7, Section 7.4.2, on page 240.

Worked example 8.6

The rotational energy levels $F(J)$ of a rigid diatomic molecule are given by the expression

$$F(J) = BJ(J + 1)$$

Each energy level has a degeneracy of $(2J+1)$; in other words, there are $(2J+1)$ levels having each energy value. The population of each level of energy $F(J)$ is proportional to

$$(2J + 1)\exp\left(-\frac{F(J)}{kT}\right).$$

where J is the rotational quantum number (taking values 0, 1, 2, ...) and B is the rotational constant. What are the three lowest rotational energy levels in terms of B?

Chemical background

This problem shows that the rotational energy levels in a diatomic molecule are not evenly spaced. However, if we consider the differences between levels, then we do see an even spacing, and this is apparent in rotational spectra.

Solution to worked example 8.6

To obtain the three lowest energy levels, we need to substitute the three lowest values of J into the given expression. We then have

$$F(0) = (B \times 0)(0 + 1) = 0$$

$$F(1) = (B \times 1)(1 + 1) = 2B$$

$$F(2) = (B \times 2)(2 + 1) = 6B$$

as shown in Figure 8.5.

Figure 8.5   Spacing of rotational energy levels.

Worked example 8.7

The rotational energy levels $F(J)$ *of* a rigid diatomic molecule are given by the expression

$$F(J) = BJ(J + 1)$$

where J is the rotational quantum number and B is the rotational constant. Obtain an expression for the energy of a rotational transition, given that such transitions are subject to the selection rule $\Delta J = \pm 1$.

Chemical background

The rotational constant B can be calculated from the equation

$$B = \frac{h^2}{8\pi^2 I}$$

where h is Planck's constant and I is the moment of inertia. The selection rule is obtained from the application of the Schrödinger equation.

A rotational spectrum will only be seen in the case of heteronuclear diatomic molecules. These will be able to interact with the radiation as their dipole moment changes during the rotation. No such spectrum would be seen for Cl_2, but it would for HCl.

Solution to worked example 8.7

To calculate the energy of a transition, we need to obtain expressions for the energies of two adjacent levels. We choose these to have rotational quantum numbers J and $J + 1$, respectively. We then have

$$F(J) = BJ(J + 1)$$

as given above.

To obtain $F(J + 1)$, we need to replace each occurrence of J in the above equation by $J + 1$. This gives

$$F(J + 1) = B(J + 1)((J + 1) + 1)$$

$$= B(J + 1)(J + 2)$$

We multiply the two brackets together by considering the four combinations of terms to obtain

$$F(J + 1) = B(J^2 + 2J + J + 2)$$

$$= B(J^2 + 3J + 2)$$

The energy difference $\Delta F(J)$ is given by

$$\Delta F(J) = F(J + 1) - F(J)$$

$$= B(J^2 + 3J + 2) - BJ(J + 1)$$

$$= B(J^2 + 3J + 2 - J^2 - J)$$

$$= B(2J + 2)$$

We can remove the common factor of 2 from within the bracket to leave

$$\Delta F(J) = 2B(J + 1)$$

8.6   Vibrational spectroscopy

The starting point for a study of vibrational spectroscopy is usually the model of a simple harmonic oscillator, a system that is often treated in some detail in courses on quantum mechanics. Any energy possessed by this system over and above its equilibrium value is due to extension or compression of the bond. The vibrational frequency, ω_e, of this system is given by the formula

$$\omega_e = \frac{1}{2\pi}\sqrt{\frac{k}{\mu}}$$

where k is the force constant and μ is a quantity known as the reduced mass of the system, which is related to the masses of the atoms forming the bonds.

Worked example 8.8

The vibrational energy $G(v)$ of the harmonic oscillator is given by the equation

$$G(v) = \left(v + \frac{1}{2}\right)\omega_e$$

where v is the vibrational quantum number (having values 0, 1, 2, …). Calculate the energy of the vibrational level having $v = 2$ for the HCl molecule. Its force constant is 516 N m^{-1}, and its reduced mass is given by the equation

$$\mu = \frac{m_H m_{Cl}}{m_H + m_{Cl}}$$

where m represents the mass of the specified atom.

Chemical background

In practice, there will be very few molecules in this vibrational level, so it will not contribute significantly to the vibrational spectrum.

Solution to worked example 8.8

For 1 mol of HCl we would have

$$\mu = \frac{35.453\,\text{g mol}^{-1} \times 1.0079\,\text{g mol}^{-1}}{35.453\,\text{g mol}^{-1} + 1.0079\,\text{g mol}^{-1}}$$

$$= \frac{35.733\,\text{g mol}^{-1}}{36.461}$$

$$= 0.98003\,\text{g mol}^{-1}$$

and for one molecule we need to divide by Avogadro's constant:

$$\mu = \frac{0.9800\,\text{g mol}^{-1}}{6.022 \times 10^{23}\,\text{mol}^{-1}}$$

$$= \frac{0.1627\,\text{g mol}^{-1}}{10^{23}\,\text{mol}^{-1}}$$

$$= 1.627 \times 10^{-24}\,\text{g}$$

$$= 1.627 \times 10^{-27}\,\text{kg}$$

so, from the formula above, the fundamental frequency can be calculated as

$$\varpi_e = \frac{1}{2\pi}\sqrt{\frac{k}{\mu}}$$

$$= \frac{1}{2\pi}\sqrt{\frac{516\,\text{N m}^{-1}}{1.627 \times 10^{-27}\,\text{kg}}}$$

Because 1 N = 1 kg m s^{-2}, this becomes

One way of remembering the correct formula for calculating the reduced mass is to consider the units. In the formula here, we have units of mass squared divided by units of mass; the reciprocal would clearly give incorrect units.

$$\omega_e = \frac{1}{2\pi} \sqrt{\frac{516\,\mathrm{kg\,m\,s^{-2}\,m^{-1}}}{1.627 \times 10^{-27}\,\mathrm{kg}}}$$

Cancelling units (kg top and bottom, and m and m⁻¹) then gives

$$\omega_e = \frac{1}{2\pi} \sqrt{\frac{516\,\mathrm{s^{-2}}}{1.627 \times 10^{-27}}}$$

$$= \frac{\sqrt{3.171 \times 10^{29}\,\mathrm{s^{-2}}}}{2\pi}$$

$$= \frac{5.631 \times 10^{14}\,\mathrm{s^{-1}}}{2\pi}$$

$$= 8.962 \times 10^{13}\,\mathrm{s^{-1}}$$

We now need to substitute the value of $v = 2$ into the expression for the vibrational energy. This gives us

$$G(2) = \left(2 + \frac{1}{2} \right) \omega_e$$

$$= 2.5 \times 8.962 \times 10^{13}\ \mathrm{s^{-1}}$$

$$= 2.241 \times 10^{14}\ \mathrm{s^{-1}}$$

$$= 2.241 \times 10^{14}\ \mathrm{Hz}$$

because $1\ \mathrm{Hz} = 1\ \mathrm{s^{-1}}$.

Use the √ calculator key to determine the square root.

This answer is easily transformed into other quantities by the use of appropriate conversion factors. Multiplication by Planck's constant h will give energy (in J), and division by the velocity of light c will give wavenumber (in m⁻¹, easily converted to cm⁻¹).

Worked example 8.9

The vibrational energy $G(v)$ of the harmonic oscillator is given by the equation

$$G(v) = \left(v + \frac{1}{2} \right) \omega_e$$

where v is the vibrational quantum number. Given that the selection rule for a vibrational transition is $\Delta v = \pm 1$, obtain an expression for the energy associated with such a transition.

The harmonic oscillator approximation is valid for vibrations that displace the bond length by up to about 10% of its equilibrium value. Above that, it is better to use the model of an anharmonic oscillator, as used in problem 6 at the end of this chapter.

Chemical background

The energy associated with this transition turns out to be a constant value, equal to the vibrational frequency of the system. This is because the vibrational energy levels are equally spaced and the vibrating molecule will only absorb energy of the same frequency as its own natural vibrations.

Solution to worked example 8.9

As in the rotational case in worked example 8.4, we need to set up expressions for the energy of two adjacent levels. We have from the question,

$$G(v) = \left(v + \frac{1}{2}\right)\omega_e$$

Replacing every occurrence of v by $v + 1$ gives

$$G(v+1) = \left(\left(v+1\right) + \frac{1}{2}\right)\omega_e$$

$$= \left(v + 1 + \frac{1}{2}\right)\omega_e$$

$$= \left(v + \frac{3}{2}\right)\omega_e$$

The energy $\Delta G(v)$ associated with this vibrational transition will now be given by

$$\Delta G(v) = G(v+1) - G(v)$$

$$= \left(v + \frac{3}{2}\right)\omega_e - \left(v + \frac{1}{2}\right)\omega_e$$

$$= \left(v + \frac{3}{2} - v - \frac{1}{2}\right)\omega_e$$

$$= \omega_e$$

The spacing of vibrational energy levels is shown in Figure 8.6.

Figure 8.6   Spacing of vibrational energy levels.

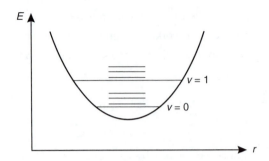

Figure 8.7   Vibrational and rotational energy levels on the same scale.

8.7   Rotation–vibration spectroscopy

It is reasonable to suppose that a molecule will be undergoing both rotational and vibrational motion at a given time, so it is not surprising that it is possible to investigate the coupling of these motions by means of spectroscopic techniques. We assume that it is possible to separate the vibrational and rotational motion, and it is therefore possible to add the contributions from each to obtain the vibration–rotation energy. Figure 8.7 shows the rotational levels superimposed on the vibrational levels for a molecule.

Worked example 8.10

The energy $E(v, J)$ due to combined vibration and rotation is given by the equation

$$E(v,J) = BJ(J+1) + \left(v + \frac{1}{2} \right) \omega_e$$

where v and J are the vibrational and rotational quantum numbers, respectively. Given the selection rules $\Delta v = \pm 1$ and $\Delta J = \pm 1$, obtain an expression for the energy due to simultaneous vibrational and rotational transitions.

Chemical background

It is important to realize that the separation of the rotational levels is much less than that of the vibrational levels. Typical values are 0.05 kJ mol^{-1} for rotational levels and 10 kJ mol^{-1} for vibrational levels. These compare with a separation of the order of 500 kJ mol^{-1} for electronic levels.

The selection rules given show that it is only possible for a vibrational change to occur if a rotational change takes place simultaneously. It is possible to have $\Delta v = 0$, which represents the pure rotational change we met in worked example 8.4, but it is not possible to have $\Delta J = 0$ representing a vibrational change alone.

Solution to worked example 8.10

We proceed as in the previous problems by considering the difference between two energy levels, remembering this time that we need to deal with two simultaneous changes. The notation usually employed is that representing a transition from a lower vibrational level v'' and rotational level J'' to an upper vibrational level v' and corresponding rotational level J'. We then have

$$\Delta E(v,J) = E(v',J') - E(v'',J'')$$

$$= BJ'(J'+1) + \left(v' + \frac{1}{2}\right)\omega_e - BJ''(J''+1) - \left(v'' + \frac{1}{2}\right)\omega_e$$

$$= BJ'(J'+1) - BJ''(J''+1) + (v'-v'')\omega_e$$

Because we have the selection rule $\Delta v = \pm 1$, it follows that

$$v' - v'' = 1$$

because v' is the higher level, and so,

$$\Delta E(v, J) = BJ(J'+1) - BJ''(J''+1) + \omega_e$$

When we move to the higher vibrational level, it is possible for the change in rotational quantum number to be either positive or negative. If

$$J' = J'' + 1$$

we have

$$\Delta E(J') = BJ'(J'+1) - B(J'-1)J' + \omega_e$$

$$= BJ'^2 + BJ' - BJ'^2 + BJ' + \omega_e$$

$$= \omega_e + 2\,BJ'$$

Alternatively, if

$$J' = J'' - 1$$

then

$$\Delta E(J'') = B(J''-1)J'' - BJ''(J''+1) + \omega_e$$

$$= BJ''^2 - BJ'' - BJ''^2 - BJ'' + \omega_e$$

$$= \omega_e - 2BJ''$$

Note that either of these expressions could be obtained using J' or J''.

8.8   Nuclear magnetic resonance spectroscopy

Chemical shifts can be measured on the τ scale, on which tetramethylsilane has a value of 10. The chemical shifts of most groups of interest then fall between 0 and 10, with a phenyl hydrogen having a value of 2.7, for example.

This form of spectroscopy is one of the most powerful analytical tools used by organic chemists. Although spectra for complicated molecules can be quite difficult to unravel, the underlying mathematics are quite simple to understand. Such spectra are basically due to the splitting of the proton spin energy levels by a magnetic field. Peaks are classified by their chemical shifts (measured in parts per million, ppm) relative to the standard tetramethylsilane, $SiMe_4$.

8.8.1   Pascal's triangle

If we wish to expand the expression $(x + 2)^2$, we could rewrite it as

$$(x + 2)^2 = (x + 2)(x + 2)$$

and multiply out the brackets to give

$$(x + 2)^2 = x(x + 2) + 2(x + 2)$$
$$= x^2 + 2x + 2x + 4$$
$$= x^2 + 4x + 4$$

Similarly, $(x + 2)^3$ could be calculated as

$$(x + 2)^3 = (x + 2)(x + 2)^2$$
$$= (x + 2)(x^2 + 4x + 4)$$
$$= x (x^2 + 4x + 4) + 2(x^2 + 4x + 4)$$
$$= x^3 + 4x^2 + 4x + 2x^2 + 8x + 8$$
$$= x^3 + 6x^2 + 12x + 8$$

In a similar fashion we could obtain expressions for $(x + 2)^4$ and higher powers, but this is obviously going to become more and more difficult as the power increases.

Fortunately, such step-by-step expansion is not necessary if we use a device known as *Pascal's triangle*. This is shown in Figure 8.8. You will be relieved to know that this does not need to be memorized, as it can be set up using a few simple rules that are easily remembered.

Figure 8.8   Pascal's triangle.

We start off on the top line with the number 1.

1

This will actually be of more use to us if we write it with zeros on either side as

0 1 0

To obtain the second line, we now add each pair of numbers in the line above and write the result in the space underneath them. This gives

0 1 0

+ +

1 1

We now add the zeros on either side of this second line:

0 1 0

0 1 1 0

To obtain the third line, we repeat the process of adding each pair of numbers in the second line and writing the result in the space underneath. We then have

0 1 0

0 1 1 0

+ + +

0 1 2 1 0

The rest of Pascal's triangle can be built up in the same way.

How does this help us to expand brackets to a specified power? From the examples given, we can see that when a bracket such as $(x + 2)$ is raised to some power, we obtain a polynomial expression. These were discussed in Chapter 3, Section 3.4.2, on page 61, where we saw that each term in x is multiplied by a constant called the coefficient. For example, in a term such as $3x^4$, we say that the coefficient of x^4 is 3. It is these coefficients that can be determined by using Pascal's triangle.

From Figure 8.8, we see that the first line of Pascal's triangle gives us the coefficient of a zero-order expansion (power 0), the next line a first-order expansion (power 1), and so on. Therefore, to evaluate $(x + 2)^2$, we need the coefficients from the third line, which are 1, 2, 1. As we saw earlier, we would expect this expansion to give us terms in x^2, x, and a constant term, so we have

coefficient of $x^2 = 1$

coefficient of $x = 2$

coefficient of constant $= 1$

In fact, this expansion can be expressed as

$$(x + 2)^2 = 1 \times x^2 \times 2^0 + 2 \times x^1 \times 2^1 + 1 \times x^0 \times 2^2$$

where each term comprises the appropriate coefficient from Pascal's triangle, a term in x and a term in 2. Notice that as the power of x decreases, that of 2 increases, but the two powers added together are always equal to 2. If we remember that any number raised to a zero power is 1, and any number raised to power 1 is itself, this expression simplifies to give

$$(x + 2)^2 = 1 \times x^2 \times 1 + 2 \times x \times 2 + 1 \times 1 \times 4$$

$$= x^2 + 4x + 4$$

which is, of course, the result we obtained before.

Let us now consider an example which would be much more difficult to evaluate without Pascal's triangle, $(2x + 3)^4$. From Figure 8.8 we see that the appropriate coefficients for an expansion of degree 4 are

$$1\ 4\ 6\ 4\ 1$$

This time we need to consider terms in $2x$ and 3, so we have

$$(2x + 3)^4 = 1 \times (2x)^4 \times 3^0 + 4 \times (2x)^3 \times 3^1$$

$$+ 6 \times (2x)^2 \times 3^2 + 4 \times (2x)^1 \times 3^3 + 1 \times (2x)^0 \times 3^4$$

Notice that when we raise a bracket to the fourth power, we have five terms. The power of $(2x)$ decreases, whereas that of 3 increases, and the sum of these powers is always 4. It is important to realize that, for example,

$$(2x)^4 = 2^4 x^4 = 16x^4$$

so the power is applied to *both* quantities within the bracket. The expansion can then be simplified to

$$(2x + 3)^4 = 1 \times 16x^4 \times 1 + 4 \times 8x^3 \times 3 + 6 \times 4x^2 \times 9$$

$$+ 4 \times 2x \times 9 + 1 \times 1 \times 81$$

$$= 16x^4 + 96x^3 + 216x^2 + 72x + 81$$

Worked example 8.11

The peaks due to each group of protons in a molecule are split into $n + 1$ by an adjacent group of n protons, the intensity of these peaks being given by the coefficients of Pascal's triangle. What is the distribution of peaks within the CH_2 and CH_3 signals of ethanol, ignoring any interaction with the OH group?

Chemical background

The reason for ignoring any interaction with the OH group is that its proton undergoes rapid exchange with other molecules due to the presence of small amounts of acid, and so it is unable to interact with protons in the other groups of the molecule.

Figure 8.9   Nuclear magnetic resonance spectrum of ethanol: (a) triplet due to CH_3 group; and (b) quartet due to CH_2 group.

Solution to worked example 8.11

The signal due to the CH_3 group will be split into three (2 + 1) peaks by the CH_2 group, and from Pascal's triangle these will be in the ratio 1 : 2 : 1. The signal due to the CH_2 group will be split into four (3 + 1) peaks by the CH_3 group, in the ratio 1 : 3 : 3 : 1. These peaks are shown in Figure 8.9.

Nuclear magnetic resonance spectra are most commonly measured for the ^{1}H nucleus. Other nuclei that can be used include ^{19}F, ^{31}P, and, since the advent of Fourier transform methods, [13]C.

8.9   Fourier transform spectroscopy

Fourier transforms are used in both infrared and nuclear magnetic resonance spectroscopy. Essentially, they allow conversion between waveforms and spectra, allowing the analysis of a complicated spectrum to give its components in terms of individual frequencies. The spectrometers used contain a built-in computer that performs the necessary integrations.

A further use for Fourier transforms is in x-ray crystallography.

8.9.1   Introduction to Fourier transforms

If two functions $f(x)$ and $g(y)$ are related by the equations

$$f(x) = \int_{-\infty}^{\infty} g(y)e^{ixy}dy$$

and

$$g(y) = \frac{1}{2\pi} \int_{-\infty}^{\infty} f(x)e^{-ixy}dx$$

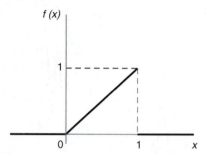

Figure 8.10   The function $f(x)$ defined in Section 8.9.1.

then we say that $g(y)$ is the Fourier transform of $f(x)$ and that $f(x)$ is the inverse Fourier transform of $g(y)$.

When determining Fourier transforms, we frequently have discontinuous functions; in these cases the form of the function varies according to the value of the variable. For example, a function $f(x)$ could be defined as

$$f(x) = \begin{cases} 0 \text{ for } x < 0 \\ x \text{ for } 0 < x < 1 \\ 0 \text{ for } x > 1 \end{cases}$$

as shown in Figure 8.10.

The Fourier transform of $f(x)$ is then defined by

$$g(y) = \frac{1}{2\pi} \int_0^1 xe^{-ixy} dx$$

because the integrals for $x < 0$ and $x > 1$ will be zero. Because the remaining integral is with respect to the variable x, we treat y as constant and therefore need to use the standard integral

$$\int xe^{ax} dx = \frac{e^{ax}}{a^2}(ax - 1) + C$$

Consequently,

$$g(y) = \frac{1}{2\pi} \left[\frac{e^{-ixy}}{(-iy)^2}(-ixy - 1) \right]_0^1$$

This can be simplified because

$$(-iy)^2 = i^2 y^2$$

$$= -y^2$$

so that

$$g(y) = \frac{1}{2\pi} \left[\frac{e^{-ixy}}{-y^2}(-ixy - 1) \right]_0^1$$

$$= \frac{1}{2\pi} \left[\frac{e^{-ixy}}{y^2}(1 + ixy) \right]_0^1$$

Now, substituting for the limits in x,

$$g(y) = \frac{1}{2\pi} \left\{ \left[\frac{e^{-iy}}{y^2}(1 + iy) \right] - \left[\frac{e^0}{y^2} \right] \right\}$$

$$= \frac{1}{2\pi} \left[\frac{e^{-iy}}{y^2}(1 + iy) - \frac{1}{y^2} \right]$$

$$= \frac{1}{2\pi y^2} \left[e^{-iy} + iye^{-iy} - 1 \right]$$

This can be further simplified to extract the real and imaginary parts of $g(y)$.

Worked example 8.12

In Fourier transform spectroscopy, radiation can be applied to a sample in terms of a pulse. This can be described in terms of a function of time $f(t)$; for example,

$$f(t) = \begin{cases} 0 \text{ for } t < 1 \\ 2 \text{ for } 1 < t < 2 \\ 0 \text{ for } t > 2 \end{cases}$$

as shown in Figure 8.11. Determine the Fourier transform $g(\omega)$ in terms of frequency ω.

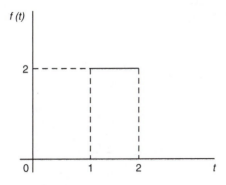

Figure 8.11 The function $f(t)$ defined in worked example 8.12.

Chemical background

Readable introductions to the use of Fourier transforms in chemistry can be found in *Chemistry in Britain*, May 1987, pp. 440–446, and *Journal of Chemical Education* **64**, A228–A233, 1987.

This is a very simplistic representation of the application of radiation, but it does allow us to illustrate the use of the Fourier transform to move from a time domain to a frequency domain. A more realistic shape for the pulse of radiation would be a Gaussian distribution; in that case the resulting Fourier transform is also a Gaussian-shaped normal distribution.

Solution to worked example 8.12

The equations for the Fourier transform in terms of the variables t and ω become

$$f(t) = \int_{-\infty}^{\infty} g(\omega) e^{it\omega} d\omega$$

and

$$g(\omega) = \frac{1}{2\pi} \int_{-\infty}^{\infty} f(t) e^{-it\omega} dt$$

Consequently,

$$g(\omega) = \frac{1}{2\pi} \int_{1}^{2} 2 e^{-it\omega} dt$$

This is a straightforward integration of the form

$$\int e^{ax} dx = \frac{e^{ax}}{a} + C$$

so that

$$g(\omega) = \frac{1}{2\pi} \left[2 \frac{e^{-it\omega}}{-it\omega} \right]_{1}^{2}$$

$$= \frac{1}{2\pi} \left[\frac{2e^{-2i\omega}}{-2i\omega} - \frac{2e^{-i\omega}}{-i\omega} \right]$$

$$= -\frac{1}{\pi i \omega} \left[\frac{e^{-2i\omega}}{2} - e^{-i\omega} \right]$$

Exercises

1. Given the values

i	1	2	3	4	5
a_i	2.3	1.8	0.6	4.2	5.1
x_i	-0.16	0.19	-0.24	0.54	0.88

   Determine the values of $\sum_i a_i x_i$ and $\sum_i a_i x_i^2$.

2. If $s = \dfrac{1}{t}$ and $t = \dfrac{u}{v}$, calculate s if $u = 41.6$ and $v = 15.8$.

3. If z is defined by

$$z = \frac{xy}{x + y}$$

   calculate z if $x = 14.9$ and $y = -3.2$.

4. Identify the type of proportion between the appropriate functions of x and y in the relationships:

   (a) $x \ln y = K$

   (b) $\dfrac{x^2}{y} = K$

   (c) $xe^{-y} = K$

   where K is a constant.

5. Identify the proportionality constant when y is written as a function of x for the relationships

   (a) $ax = by$

   (b) $\left(\dfrac{x}{a}\right)^2 = \left(\dfrac{b}{y}\right)$

   (c) $\dfrac{y}{\ln x} = \dfrac{a}{b}$

6. Write each of these expressions as a single logarithm:

   (a) $\ln x^3 - \ln x^2$

   (b) $\ln xy + \ln x^2$

   (c) $\ln\left(\dfrac{x}{y}\right) - \ln(xy)$

7. Write each of these expressions as a sum or difference of logarithms:

   (a) $\ln\sqrt{\dfrac{x+1}{x-1}}$

   (b) $\ln 5x^4$

   (c) $\ln\left(\dfrac{3x^2y}{z}\right)$

8. Simplify the expressions:

   (a) $\log(4x+2) - \log(2x+1)$

   (b) $\log(x^2 - 2x) - \log(x-2)$

   (c) $\log 5x^3 - \log x$

9. The terms in a series are defined by the expression

   $$f(n) = 3n^3 + 4n^2 - 2n + 1$$

   What is the value of $f(2) - f(1)$?

10. The terms in a series are defined by the expression

   $$g(N) = N^2 + 2N$$

   Obtain an expression for $g(N+1) - g(N)$.

11. Verify that the sequence

   $$1, \frac{3}{2}, \frac{7}{4}, \frac{15}{8}, \frac{31}{16}$$

   can be represented as

   $$f(n) = 2 - \frac{1}{2^{n-1}}$$

   What is the limit of this sequence as n tends to infinity?

12. The sine function can be represented by a series in x by means of the equation:

$$\sin x = a + bx + cx^2$$

    (a)   Set $x = 0$ to determine the value of a.
    (b)   Differentiate this equation to give an expression for $\cos x$.
    (c)   Set $x = 0$ to determine the value of b.
    (d)   Repeat the process and determine the value of c.
    (e)   Hence predict the series expression for $\sin x$.

13. Use Pascal's triangle to obtain expressions for

    (a) $(x + 3)^4$

    (b) $(2x - 4)^5$

    (c) $(3x + 2)^6$

14. Use Pascal's triangle to obtain expressions for:

    (a) $(x + y)^5$

    (b) $\left(\dfrac{3}{x} - y \right)^4$

    (c) $(2x + 5y)^4$

15. In the expansion of $(x - 3y)^4$, what is the coefficient of y^3?

16. Use Pascal's triangle to obtain an expression for $(1 - x)^6$. By setting x to 0.002, obtain the value of 0.998 to 6 significant figures *without* using a calculator.

17. Obtain a general expression for the sum of the numbers in each row n of Pascal's triangle.

18. The Fibonacci sequence

$$1, 1, 2, 3, 5, 8 \ldots$$

is formed by starting with 1 and adding the two previous numbers. It has recently been popularized by being featured in Dan Brown's *The Da Vinci Code*. Show that the Fibonacci sequence is given by the sums of the diagonals in Pascal's triangle if the triangle is written with the first number in each line at the left margin.

19. Determine the Fourier transform of

$$f(x) = \begin{cases} 0 \text{ for } x < -\pi \\ \sin 3x \text{ for } -\pi < x < \pi \\ 0 \text{ for } x > \pi \end{cases}$$

using the standard integral

$$\int e^{ax} \sin bx\, dx = \frac{e^{ax}}{a^2 + b^2}(a \sin bx - b \cos bx) + C$$

20. Determine the Fourier transform of

$$f(x) = \begin{cases} 0 \text{ for } x < 0 \\ x^2 \text{ for } 0 < x < 1 \\ 0 \text{ for } x > 1 \end{cases}$$

using the standard integral

$$\int x^n e^{ax}\, dx = \frac{1}{a}x^n e^{ax} - \frac{n}{a}\int x^{n-1} e^{ax}\, dx$$

Problems

1. Calculate the dipole and quadrupole moments for HCl, which has a bond length 1.31 Å, and calculated partial charges of $0.673e$.

2. Express a wavenumber of 2318 cm^{-1} in terms of frequency, wavelength, and energy.

3. Lambert's law states that each successive layer of thickness dx of a medium absorbs an equal fraction

$$-\frac{dI}{I}$$

of radiation of intensity I, so that

$$-\frac{dI}{I} = b\, dx$$

where b is a constant. Obtain an expression for the intensity I of light that has traveled through a distance l, assuming that $I = I_0$ when $l = 0$.

4. The rotational energy levels $F(J)$ of a nonrigid molecule are given by the formula

$$F(J) = BJ(J + 1) - DJ^2(J + 1)^2$$

where J is the rotational quantum number, B the rotational constant, and D is the centrifugal distortion constant. Obtain an expression for

$$F(J + 1) - F(J)$$

in terms of B and D.

5. Determine the energy of the $J = 1$ to $J = 2$ transition in HCl, which has a rotational constant of 10.5937 cm^{-1} and a centrifugal distortion constant of 4.1×10^{-5} cm^{-1}.

6. The vibrational energy levels $G(v)$ of an anharmonic oscillator are given by the expression

$$G(v) = \left(v + \frac{1}{2}\right)\omega_e - \left(v + \frac{1}{2}\right)^2 \omega_e x_e$$

where v is the vibrational quantum number, ω_e is the vibrational frequency and x_e is the anharmonicity constant. Obtain an expression for $G(1) - G(0)$, which is the most likely vibrational transition because higher levels will only be very sparsely populated.

7. Determine the wavelength of the $v = 0$ to $v = 1$ transition for HCl, which has a fundamental vibrational frequency of 2988.9 cm^{-1} and an anharmonicity constant of 1.540×10^{-2}.

8. The wavenumbers $\tilde{v}$ of lines in the rotation–vibration spectrum of carbon monoxide form a series defined by

$$\frac{\tilde{v}(m)}{\text{cm}^{-1}} = 2143.28 + 3.813m - 0.0175m^2$$

where $m = ..., -2, -1, 0, 1, 2, ...$.

Obtain an expression for $\tilde{v}(m + 1) - \tilde{v}(m)$ and show that for positive values of m the line spacing decreases, whereas for negative values it increases with increasing magnitude of m.

9. The peaks due to each group of protons in a molecule are split into $n + 1$ by an adjacent group of n protons, the intensities of these peaks being given by the coefficients of Pascal's triangle. Describe the appearance of the nuclear magnetic resonance spectrum of isopropyl iodide, $(CH_3)_2CHI$.

10. The exponential decay function, exemplified by

$$f(t) = \begin{cases} 0 \text{ for } t < 0 \\ e^{-at} \text{ for } 1 < t < 2 \\ 0 \text{ for } t > 2 \end{cases}$$

is important in Fourier transform spectroscopy. Determine its Fourier transform $g(\omega)$.

Statistical mechanics 9

Thermodynamics only deals with a macroscopic picture of matter, and its principles have all been derived by observing experimental behaviour. In contrast, quantum mechanics is concerned with the microscopic view and it starts from a more theoretical point, even though many of its results do explain experimental behaviour. Invariably, in quantum mechanics we are considering only single atoms or molecules.

The topic of statistical mechanics provides a link between these subjects. We start with the microscopic view of matter and apply statistical techniques to a large number of chemical entities in order to reproduce the macroscopic functions of thermodynamics. This requires a few mathematical techniques we have not yet come across, as well as some of those met in earlier chapters.

9.1 Molecular energy distributions

Much of statistical mechanics considers the distribution of molecules across energy levels. As you might expect, when we are dealing with large numbers of molecules, some of the ideas we need to use may not be immediately obvious. To overcome this problem, we can often learn much about the principles we need to apply by considering simple systems.

Worked example 9.1

In how many ways can 4 molecules be distributed across 3 energy levels, such that 2 are in the first level and 1 is in each of the higher levels?

Chemical background

Specifying the distribution of molecules in this way is equivalent to specifying the state of the system. The more ways there are of achieving a given state, the more probable is the chance that the state will exist. In practice, it is found that one state of system predominates, and this can aid the analysis considerably.

Solution to worked example 9.1

The arrangement of molecules within energy levels is illustrated in Figure 9.1. If we denote the four molecules as a, b, c, and d, we can construct a table of possible arrangements as shown:

d	c	d	b	c	b	d	a	c	a	b	a
c	d	b	d	b	c	a	d	a	c	a	b
a, b	a, b	a, c	a, c	a, d	a, d	b, c	b, c	b, d	b, d	c, d	c, d

Figure 9.1   The distribution of molecules described in worked example 9.1.

This can be constructed by first considering each pair of molecules that can occupy the lowest level, viz., a and b, a and c, a and d, b and c, b and d, and c and d. For each of these, the remaining 2 molecules can then occupy the upper levels in each of two ways. There are consequently 12 ways in which this can be achieved.

9.2   Configurations

Strictly speaking, we are dealing with instantaneous configurations, which will be constantly changing.

The number of ways in which a specified number of particles can be arranged in each energy level is known as the number of complexions. For example, we might wish to specify 2 molecules in the lowest energy level, 3 in the next level, and 4 in the level above. Such an arrangement can only be achieved in a certain number of ways starting with 9 molecules.

9.2.1   Factorials

A typical calculator key sequence for calculating a

factorial would be

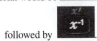

followed by   .

A factorial is a number written with an exclamation mark (!) immediately after it, such as 2!, 45!, or 8! This strange-looking notation has a simple meaning, as illustrated by the following examples:

$$5! = 5 \times 4 \times 3 \times 2 \times 1 = 120$$

$$9! = 9 \times 8 \times 7 \times 6 \times 5 \times 4 \times 3 \times 2 \times 1 = 362\,880$$

In general, for a number n we have

$$n! = n \times (n-1) \times (n-2) \times (n-3) \ldots 3 \times 2 \times 1$$

Worked example 9.2
The number of complexions Ω is given by the equation

$$\Omega = \frac{N!}{\prod_i n_i!}$$

where N is the total number of particles and n_i is the number of particles in energy level i. The symbol $\prod$ denotes that we calculate the product

Figure 9.2   The distribution of molecules as described in worked example 9.2.

of the quantities specified. What is the number of complexions if 3 molecules are in level 1, 2 molecules in level 2, and 1 molecule in level 3, as shown in Figure 9.2?

Chemical background

The arrangement of molecules between these energy levels is known as a state and will be constantly changing. Normally, the broadest distribution of the molecules across the energy levels will predominate. The expression for the number of complexions provides a starting point for the derivation of the equation that describes the Boltzmann distribution law.

Solution to worked example 9.2

The first stage is to write out the quantities we have been given in the question using the notation of the given equation. If there are n_i molecules in level i, we can write

$$n_1 = 3, n_2 = 2, n_3 = 1$$

and the total number N of molecules is then given as

$$N = n_1 + n_2 + n_3$$
$$= 3 + 2 + 1$$
$$= 6$$

We can now calculate $N!$ as

$$N! = 6!$$
$$= 6 \times 5 \times 4 \times 3 \times 2 \times 1$$
$$= 720$$

The other factorials required are

$$n_1! = 3! = 3 \times 2 \times 1 = 6$$
$$n_2! = 2! = 2 \times 1 = 2$$

and

$$n_3! = 1! = 1$$

We can now substitute into our equation to give

$$\Omega = \frac{N!}{\prod_i n_i!}$$

$$= \frac{N!}{n_1! n_2! n_3!}$$

$$= \frac{6!}{3!2!1!}$$

$$= \frac{720}{6 \times 2 \times 1}$$

$$= \frac{720}{12}$$

$$= 60$$

This makes use of the

calculator button sequence.

There is, however, a slightly neater way of performing this calculation. It will be shown here because the technique can be of use when we are dealing with symbols rather than actual numbers.

Because

$$6! = 6 \times 5 \times 4 \times 3 \times 2 \times 1$$

and

$$3! = 3 \times 2 \times 1$$

we can write

$$6! = 6 \times 5 \times 4 \times 3!$$

If we now substitute into the equation for Ω, we obtain

$$\Omega = \frac{6!}{3!2!1!}$$

$$= \frac{6 \times 5 \times 4 \times 3!}{3!2!1!}$$

Cancelling 3! top and bottom, and writing 2! as 2, and 1! as 1 gives

$$\Omega = \frac{6 \times 5 \times 4}{2}$$

$$= 6 \times 5 \times 2$$

$$= 60$$

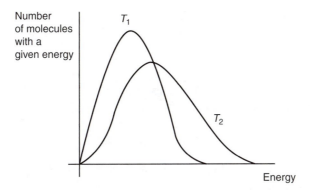

Figure 9.3   The Boltzmann distribution with $T_2 > T_1$.

This second method makes the calculation easier when a calculator is not available.

9.3   The Boltzmann equation

The Boltzmann distribution law allows us to specify the distribution of molecules among the energy levels. An increase in temperature results in an increase in molecular energy, as shown in Figure 9.3. Therefore, we would expect such a function to depend upon temperature.

It is not always appreciated that the Boltzmann distribution is given in terms of general energy levels, although the Maxwell–Boltzmann distribution relates only to kinetic energy and molecular speeds.

The essential process in its determination is to maximize the number of complexions Ω with respect to the number of particles in each level n_i. If Ω is a maximum, it follows that $\ln \Omega$ will also be a maximum, so that the derivative that needs to be calculated and set equal to zero is

$$\frac{\partial \ln \Omega}{\partial n_i}$$

9.3.1   Differentiation of logarithms

This topic was covered in Chapter 5, Section 5.4.2, on page 149. The important result is

$$\frac{d \ln x}{dx} = \frac{1}{x}$$

9.3.2   Differentiation of products

This subject was described in Chapter 7, Section 7.6.1, on page 252, where we saw that if $f(x) = u(x) \, v(x)$, then

$$\frac{df(x)}{dx} = v(x)\frac{du(x)}{dx} + u(x)\frac{dv(x)}{dx}$$

Worked example 9.3

Stirling's approximation states that

$$\ln n! = n \ln n - n$$

for large values of n. In the derivation of the Boltzmann equation, this leads to the equation

$$\frac{\partial \ln \Omega}{\partial n_i} = -\frac{\partial}{\partial n_i}\left(n_i \ln n_i - n_i\right)$$

Obtain an expression for the derivative on the right-hand side of this equation.

Chemical background

Stirling's approximation is very useful when we are dealing with large numbers, as is the case in statistical mechanics. Consider how difficult it would be to calculate the factorial of Avogadro's number. If you try it on your calculator, you will see that it cannot be done, but it is straightforward using the equation above.

Solution to worked example 9.3

The expression to be differentiated consists of two terms that will be considered in turn. The first involves determining

$$\frac{\partial}{\partial n_i}\left(n_i \ln n_i\right)$$

Remember that the rule for differentiating a product is to differentiate the first function and multiply by the second, then differentiate the second function and multiply by the first. This will give us

$$\frac{\partial}{\partial n_i}\left(n_i \ln n_i\right) = \left(\ln n_i\right)\frac{\partial}{\partial n_i}\left(n_i\right) + n_i \frac{\partial}{\partial n_i}\left(\ln n_i\right)$$

Because

$$\frac{\partial}{\partial n_i}\left(n_i\right) = 1$$

and

$$\frac{\partial}{\partial n_i}\left(\ln n_i\right) = \frac{1}{n_i}$$

we obtain

$$\frac{\partial}{\partial n_i}\left(n_i \ln n_i\right) = \ln n_i + n_i \frac{1}{n_i}$$

$$= \ln n_i + 1$$

and so

$$\frac{\partial}{\partial n_i}\left(n_i \ln n_i - n_i\right) = -\left(\ln n_i + 1 - 1\right)$$

$$= -\ln n_i$$

Worked example 9.4

The Boltzmann equation for the number of particles n_i in an energy level i having energy ε_i is

$$\frac{n_i}{N} = \frac{\exp\left(-\dfrac{\varepsilon_i}{kT}\right)}{\displaystyle\sum_i \exp\left(-\dfrac{\varepsilon_i}{kT}\right)}$$

where N is the total number of particles and k is the Boltzmann constant. Obtain an expression for the number of particles in level i relative to that in level j.

Chemical background

The quantity

$$\sum_i \exp\left(-\frac{\varepsilon_i}{kT}\right)$$

in this expression is known as the partition function and is usually given the symbol θ. This quantity is of great importance in statistical mechanics. We often calculate the partition functions for translation, vibration, rotation, and electrons separately.

Solution to worked example 9.4

We can use the equation as given:

$$\frac{n_i}{N} = \frac{\exp\left(-\frac{\varepsilon_i}{kT}\right)}{\sum\limits_{i} \exp\left(-\frac{\varepsilon_i}{kT}\right)}$$

for level i, and rewrite it to give

$$\frac{n_j}{N} = \frac{\exp\left(-\frac{\varepsilon_j}{kT}\right)}{\sum\limits_{i} \exp\left(-\frac{\varepsilon_i}{kT}\right)}$$

for level j. Notice that we are still considering energy levels i in the second expression; it really does not matter what symbol we use as long as that on the summation sign Σ corresponds to that inside the summation (in this case on the energy level).

When we take the ratio of n_i and n_j, we obtain

$$\frac{n_i}{n_j} = \frac{\exp\left(-\frac{\varepsilon_i}{kT}\right)}{\exp\left(-\frac{\varepsilon_j}{kT}\right)}$$

because the denominator of the defining equations is the same for n_i and n_j. To simplify this expression, we need to use the fact that

$$\frac{a^x}{a^y} = a^{x-y}$$

so we obtain

$$\frac{n_i}{n_j} = \exp\left(-\frac{\varepsilon_i}{kT} - \left(-\frac{\varepsilon_j}{kT}\right)\right)$$

We can rewrite

$$-\varepsilon_i - (-\varepsilon_j)$$

as

$$-\varepsilon_j + \varepsilon_j$$

which is given more usually as

$$-(\varepsilon_i - \varepsilon_j)$$

and so the expression we obtain is

$$\frac{n_i}{n_j} = \exp\left(\frac{-(\varepsilon_i - \varepsilon_j)}{kT}\right)$$

9.4   The partition function

In the previous section, we met the idea of a partition function q and saw that it was defined as

$$q = \sum_i \exp\left(-\frac{\varepsilon_i}{kT}\right)$$

The partition function q defined here is actually the molecular partition function. The partition function Q for the whole system is equal to q^N, and this needs to be divided by $N!$ if the particles making up the system are indistinguishable, such as in a gas.

9.4.1   Integration by substitution

In Chapter 7, Section 7.6.2, we met the technique of integration by parts, which was used for the integration of certain types of functions that could not be treated using the simple rules we had already learned. Other functions may be integrated by making an appropriate substitution.

Suppose we need to calculate the integral

$$\int_0^1 (x + 5)^4 dx$$

One way of doing this, of course, would be to expand the bracket using Pascal's triangle as outlined in Chapter 8, Section 8.8.1, on page 290, and then to integrate each term of the polynomial individually. However, a quicker method, which is of more general use, is as follows.

If we make the substitution

$$u = x + 5$$

we can differentiate this defining equation to give

$$\frac{du}{dx} = 1$$

because the constant 5 will differentiate to zero. We can then rearrange the expression for the derivative to give

$$dx = du$$

Because this is a definite integral, we can replace our limits in the variable x with the corresponding limits in the variable u. For the lower limit, when $x = 0$, we have

$$u = x + 5$$

$$= 0 + 5$$

$$= 5$$

For the upper limit, $x = 1$, and so

$$u = x + 5$$

$$= 1 + 5$$

$$= 6$$

We are now able to work through the original expression for the integral, replacing terms relating to x with those relating to u. This involves:

- Replacing the lower limit $x = 0$ with $u = 5$
- Replacing the upper limit $x = 1$ with $u = 6$
- Replacing $(x + 5)^4$ with u^4
- Replacing dx with du

The result of these replacements is that the integral to be calculated becomes

$$\int_1^6 u^4 \, du$$

This is straightforward to calculate using the basic rule of raising the power by one and dividing by the new power. We then have

$$\int_1^6 u^4 \, du = \left[\frac{u^5}{5} \right]_5^6$$

Substituting the appropriate limits and using a calculator to evaluate the terms then gives

$$\int_1^6 u^4 du = \frac{6^5}{5} - \frac{5^5}{5}$$

$$= \frac{7776 - 3125}{5}$$

$$= \frac{4651}{5}$$

$$= 930.2$$

Worked example 9.5

The rotational partition function q_r of a diatomic molecule is given in terms of the rotational quantum number J as

$$q_r = \int_0^\infty (2J + 1) \exp\left(-\frac{J(J + 1)h^2}{8\pi^2 I k T} \right) dJ$$

where I is the moment of inertia, h is Planck's constant, k is Boltzmann's constant, and T is the absolute temperature. The upper limit on the integral sign is infinity, denoted by ∞. Calculate this integral in terms of these quantities by using the substitution $x = J(J + 1)$.

Chemical background

The fact that the rotational partition function is calculated using an integral rather than a summation is because the rotational levels are closely spaced. We are effectively treating the rotational levels as being continuous rather than discretely spaced, and under these conditions, it is appropriate to use the integral.

There is no simple algebraic expression for the rotational energy levels of an asymmetric top molecule, so the calculation of its rotational partition function is not straightforward.

The actual expression for q_r for diatomic molecules contains an additional quantity known as the symmetry number. This takes the value 1 for heteronuclear diatomics (such as HCl) and 2 for homonuclear diatomics (such as Cl_2) to allow for the fact that in the latter a rotation of 180° produces an indistinguishable orientation.

Solution to worked example 9.5

The substitution suggested in the question is clearly related to the exponential term in the expression we want to integrate. As a first step, it is useful to remove the brackets so we have

$$x = J(J+1) = J^2 + J$$

This can now be differentiated term by term using the rule that we multiply by the power and reduce the power by one. This gives

$$\frac{dx}{dJ} = 2J + 1$$

so that

$$dx = (2J + 1)dJ$$

and

$$dJ = \frac{dx}{2J + 1}$$

We can now separate variables and integrate

$$\int dJ = \int \frac{dx}{(2J + 1)}$$

It is straightforward to change the lower limit in J to one in x. When $J = 0$, we have

$$x = 0^2 + 0 = 0$$

Similarly, when $J = \infty$, it follows that $x = \infty$. It is not really correct to use infinity as a number in the above equation, but it should be apparent that the two limits are the same.

We can now substitute for each term in the original expression. It is actually easier to leave the initial $(2J + 1)$ bracket as it is, but the other changes required are:

- To replace $J(J + 1)$ by x in the exponential term

- To replace dJ by $\dfrac{dx}{2J + 1}$

It so happens that, in this example, the limits on the integral sign remain unchanged. The integral to be calculated in terms of x is therefore

$$q_r = \int_0^\infty (2J + 1) \exp\left(-\frac{xh^2}{8\pi^2 IkT} \right) \frac{dx}{(2J + 1)}$$

The terms in $2J + 1$ cancel top and bottom and we are left with

$$q_r = \int_0^\infty \exp\left(-\frac{xh^2}{8\pi^2 IkT} \right) dx$$

This expression looks more complicated than it really is, and it is actually quite straightforward to integrate. The only variable in the exponential term is x, so we can group the other quantities to form a constant A defined by

$$A = \frac{h^2}{8\pi^2 IkT}$$

We can now write a much simpler expression for the integral to be calculated, which is

$$q_r = \int_0^\infty e^{-Ax} dx$$

We saw in Chapter 7, Section 7.6.3, on page 254, that when we integrate the exponential function we obtain the same function divided by any constants. In this case the constant is $-A$, so we obtain

$$q_r = \int_0^\infty e^{-Ax} dx$$

$$= \left[\frac{e^{-Ax}}{-A} \right]_0^\infty$$

Instead of dividing by $-A$ within the brackets, this term can be brought outside to give

$$q_r = -\frac{1}{A} \left[e^{-Ax} \right]_0^\infty$$

Before we actually evaluate this expression at the specified limits, it is worth realizing that because

$$a^{-n} = \frac{1}{a^n}$$

the expression for q_r can be written as

$$q_r = -\frac{1}{A} \left[\frac{1}{e^{Ax}} \right]_0^\infty$$

If we now consider what happens at the upper limit of infinity, we see that we are taking the reciprocal of the exponential of a very large number. The exponential of a very large number is itself very large, so the reciprocal of this will be very small, and we will take it as zero. The lower limit is zero, so Ax will be zero, and we need the reciprocal of e^0. Because any number raised to the power zero is one, the reciprocal of e^0 is also one, and this is our lower limit. Substituting into the expression for q_r now gives us

$$q_r = -\frac{1}{A}\big[0-1\big]$$

$$= -\frac{1}{A}\big[-1\big]$$

The two negative signs multiply to give a positive sign, so we obtain

$$q_r = \frac{1}{A}$$

Because we chose to define A as

$$\frac{h^2}{8\pi^2 IkT}$$

we finally have

$$q_r = \frac{1}{A} = \frac{8\pi^2 IkT}{h^2}$$

9.4.2   Calculating a series using a spreadsheet

The concept of a sequence was introduced in Chapter 7, Section 7.4.2, on page 240; if we sum the consecutive terms of a sequence, we have a series. We also saw in Section 9.4 that the partition function q can be defined as

$$q = \sum_i \exp\left(-\frac{\varepsilon_i}{kT}\right)$$

where ε_i are the energy levels in a molecule, k the Boltzmann constant, and T the absolute temperature. Although we can obtain more usable expressions for the rotational and vibrational partition functions q_r and q_v, for the electronic partition function q_e we do usually sum over the actual energy levels. We then also have to take the degeneracy g_i into account, so the expression becomes

$$q = \sum_i g_i \exp\left(-\frac{\varepsilon_i}{kT}\right)$$

This type of repetitive calculation is ideally suited to a spreadsheet, where the numbers change in each term but not the form of the expression.

Worked example 9.6

In the oxygen atom, the first five electronic energy levels are given in the table. Calculate the electronic partition function at 298 K.

Level	Energy/10^{-21} J	Degeneracy
$^3P_2$	0	5
$^3P_1$	3.160	3
$^3P_0$	4.512	1
$^1D_2$	315.4	5
$^1S_0$	671.7	1

Chemical background

In this example, each energy level is labeled according to its term symbol. This has the general form

$$^{2S+1}L_J$$

The quantum number L represents the total orbital angular momentum. When L = 0 we use the symbol S, when L = 1 the letter P, and when L = 2 the letter D. The total angular momentum is represented by quantum number J, whereas S is the quantum number for spin angular momentum.

Solution to worked example 9.6

We begin by entering data for the energy and degeneracy into the spreadsheet. Put a label g_i into cell B2 and e_i into C2, with degeneracies in B4 to B8 and energies in C4 to C8. Note the way of entering a number in standard form, e.g., 3.160E-21 to represent 3.160×10^{-21}. We will also need values for k and T, so put these in cells F2 and F3, respectively, with labels in the preceding columns.

We now begin to perform the calculations by entering a formula in cell G4. This will be

$$= B4*EXP(-C4/(\$F\$2*\$F\$3))$$

Note the use of F2 and F3 to ensure that the values are always read from those actual cells rather than using relative addressing. G4 should now contain the value 5 because the exponent in this case is zero.

Figure 9.4 Spreadsheet used in worked example 9.6.

Now copy the formula from G4 into the cells through to G8. Note that the values should decrease with increasing energy. The final stage is to sum all these terms. Go to cell G10 and enter the formula

$$=SUM(G4:G8)$$

The value of 6.73 can actually be found by summing the first three terms only; a little inspection shows that the contribution higher levels make to the partition function is negligible. The final appearance of the spreadsheet is shown in Figure 9.4.

Exercises

1. If two six-sided dices are thrown, in how many ways can the numbers shown add to give an even number?
2. In how many ways could three cards belonging to the same suit be selected from a standard pack of playing cards?
3. Evaluate these quantities *without* using the factorial button on a calculator.

   (a) $\dfrac{7!}{3!}$

   (b) $\dfrac{6!}{(5!)^2}$

   (c) $\dfrac{10!}{4! \times 2!}$

4. Determine the values of

(a) $\dfrac{n!}{(n-1)!}$

(b) $\dfrac{n!}{(n-4)!}$

(c) $\dfrac{(n+2)!}{(n-1)!}$

5. Write each of these expressions in factorial form:

(a) $6 \times 5 \times 4$

(b) $\dfrac{13 \times 12}{5 \times 4 \times 3 \times 2}$

(c) $\dfrac{9 \times 8 \times 7}{6 \times 5 \times 4 \times 3}$

6. Write each of the following in factorial form:

(a) $n(n-1)(n-2)$

(b) $(n+1)n(n-1)$

(c) $(n+6)(n+5)(n+4)(n+3)$

7. Write each of these expressions as a multiple of a single factorial quantity:

(a) $6! + 7!$

(b) $2(10!) + 3(7!)$

(c) $10! - 8!$

8. Factorize the following expressions by writing them as a single factorial quantity:

(a) $n! - (n-1)!$

(b) $n! - (n-2)!$

(c) $(n+1)! + (n-1)!$

9. Show that

$$\frac{d}{dx}(\ln kx) = \frac{1}{x}$$

regardless of the value of k.

10. Obtain a general expression for

$$\frac{d}{dx}(\ln \sqrt[m]{x^n})$$

11. Differentiate the functions:

   (a) $x^2 \ln x$

   (b) $x^3 e^{2x}$

   (c) $x e^{-4x}$

with respect to x.

12. Differentiate the following functions with respect to x:

   (a) $x \sin x$

   (b) $x^2 \cos x$

   (c) $e^x \cos (3x + 1)$

13. Determine these integrals by using an appropriate substitution:

   (a) $\displaystyle\int_0^1 (2x + 1)^4 dx$

   (b) $\displaystyle\int x\sqrt{x^2 + 2}\, dx$

   (c) $\displaystyle\int_0^2 e^{2x+3}\, dx$

14. Determine the following integrals by using an appropriate substitution:

   (a) $\displaystyle\int e^x \sqrt{e^x + 2}\, dx$

   (b) $\displaystyle\int_0^2 x(x^2 + 1)^3 dx$

(c) $\displaystyle\int \frac{\ln x}{x}dx$

15. Use an appropriate substitution to determine

(a) $\displaystyle\int \frac{dx}{x-5}$

(b) $\displaystyle\int_{2}^{4} \frac{2}{3x+2}dx$

(c) $\displaystyle\int \frac{5}{2x+1}dx$

16. Evaluate the following integrals by using an appropriate substitution:

(a) $\displaystyle\int_{0}^{\frac{\pi}{2}} \sin(3x+2)dx$

(b) $\displaystyle\int \sin(x-5)dx$

(c) $\displaystyle\int_{\pi}^{2\pi} \cos(2x-1)dx$

17. Use a spreadsheet to determine the sum of the series defined by

$$f(n) = n^2$$

from $n = 1$ to $n = 10$.

18. Use a spreadsheet to determine the value of

$$\sum_{n=1}^{12} \frac{1}{n^3}$$

19. Obtain a general expression for the series

$$2 + 9 + 16 + \ldots + 107$$

and hence evaluate this sum using a spreadsheet.

20. Use a spreadsheet to evaluate the sum of the first 10 terms of the series defined by

$$f(n) = 3^{n+2}$$

for $n \geq 1$.

Problems

1. From first principles, deduce how many ways there are of arranging 5 molecules in 3 energy levels, so that there are 2 molecules in each of the 2 lowest levels.

2. The number of complexions Ω of a system is given by the expression

$$\Omega = \frac{N!}{\prod_i n_i!}$$

where N is the total number of particles and n_i is the number of particles in level i. Obtain an expression for Ω when

(a) all N particles are in different energy levels, as shown in Figure 9.5.

(b) all particles except one are in the lowest level, as shown in Figure 9.6.

3. The translational partition function q_{trans} for a monatomic gas is given by the expression

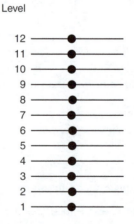

Level

12
11
10
9
8
7
6
5
4
3
2
1

Figure 9.5   All particles in different energy levels.

Level

2 ————————————●————————————

1 ●●●●●●●●●●●●●●●●●●●●

Figure 9.6   All but one particle in the lowest energy level.

$$q_{trans} = \frac{(2\pi mkT)^{\frac{3}{2}} V}{h^3}$$

where m is the mass of one molecule, k is Boltzmann's constant, T is the absolute temperature, V is the enclosing volume, and h is Planck's constant. Obtain an expression for the pressure p of the gas by using the formula

$$p = NkT \left(\frac{\partial \ln q_{trans}}{\partial V} \right)_T$$

where N is the number of molecules present.

4. (a) Obtain the equation for Stirling's approximation by writing $\ln N!$ as a sum from $n = 1$ to $n = N$, replacing it by an integral that can be evaluated using the result of Question 20 in the Exercises in Chapter 7 and neglecting appropriate terms when N is large.
   (b) Rewrite the expression for Stirling's approximation in terms of an exponential rather than a logarithmic function.
   (c) What is the error introduced by using Stirling's approximation when $N = 10$?

5. The partition function q is defined by the equation

$$q = \sum_i g_i \exp\left(-\frac{\varepsilon_i}{kT} \right)$$

where ε_i is the energy of level i, k is Boltzmann's constant, and T is the absolute temperature. Show that if q is multiplied by a positive constant, it gives q', which is the partition function for a system containing energy levels with the same spacing but a different zero.

6. It is often helpful to express the energy levels of a molecule in terms of ε_i/k as this assists computation. The electronic energy levels for the nitrogen atom are expressed in this form below:

g_i	$\dfrac{\varepsilon}{k}\Big/\mathrm{K}$
4	0
6	27658.7
4	27671.7
6	41492.4

The degeneracy g_i of each level is also given. Use this data and direct summation to show that only the ground state contributes to the electronic partition function of nitrogen at 298 K.

7. The rotational constant B of a rigid rotor is defined by the equation

$$B = \frac{h^2}{8\pi^2 I}$$

where h is Planck's constant and I the moment of inertia.

(a) The value of B for HCl is 10.5937 cm⁻¹. Convert this wavenumber to an energy value.

(b) Use the relationship

$$q_r = \frac{kT}{B}$$

to obtain the rotational partition function of HCl at 298.14 K. The absolute temperature in this equation is represented by T, and the Boltzmann constant by k.

8. Use the data from Problem 7 in Chapter 8 to determine the ratio of molecules in the $v = 0$ and $v = 1$ vibrational states at both 25°C and 100°C.

9. The Sackur–Tetrode equation gives the entropy S of an ideal gas as

$$\frac{S}{\mathrm{J\,K^{-1}\,mol^{-1}}} = -9.620 + 12.47 \ln M_r + 12.47 \ln\left(\frac{T}{\mathrm{K}}\right)$$

where M_r is the relative molecular mass and T the absolute temperature. Obtain an expression for dS/dT that describes the variation of the entropy with absolute temperature.

10. When calculating the translational partition function for a monatomic gas, it is necessary to evaluate the integral

$$\int_0^\infty \exp\left(-\frac{n_x^2 h^2}{8ma^2 kT}\right) dn_x$$

where n_x is the quantum number for direction x, a is the corresponding dimension, m is the mass of the molecule, T is the absolute temperature, h is Planck's constant, and k is Boltzmann's constant. This integral is of the general form

$$\int_0^\infty e^{-ax^2}\,dx$$

and its value can usually be obtained from tables of standard integrals. In this problem we will derive the result, however.

(a) If

$$I = \int_0^\infty e^{-ax^2}\,dx$$

show that

$$I^2 = \int_0^\infty \int_0^\infty e^{-a(x^2+y^2)}\,dA$$

where $dA = dx\,dy$.

(b) This double integral can be transformed to polar coordinates r and θ using the equations

$$r^2 = x^2 + y^2$$

$$dA = r\,dr\,dq$$

Show that this gives

$$I^2 = \int_0^\infty re^{-ar^2}\,dr \int_0^{\frac{\pi}{2}} d\theta$$

because r and θ are independent variables.

(c) Evaluate the integral

$$\int_0^\infty re^{-ar^2}\,dr$$

by using the substitution $u = r^2$ and hence show that

$$I = \frac{1}{2}\left(\frac{\pi}{a}\right)^{\frac{1}{2}}$$

Appendix A: Units

A.1 SI Prefixes

Multiplier	Name	Symbol
10^{-18}	atto	a
10^{-15}	femto	f
10^{-12}	pico	p
10^{-9}	nano	n
10^{-6}	micro	μ
10^{-3}	milli	m
10^{-2}	centi	c
10^{-1}	deci	d
10^{3}	kilo	k
10^{6}	mega	M
10^{9}	giga	G
10^{12}	tera	T

A.2 Equivalent Units

Unit	Symbol	Equivalent(s)	
Ångström	Å	10^{-10} m	
Hertz	Hz	s^{-1}	
Joule	J	N m	$kg\ m^2\ s^{-2}$
Litre	l	dm^3	$10^{-3}\ m^3$
Newton	N	$kg\ m\ s^{-2}$	
Pascal	Pa	$N\ m^{-2}$	$kg\ m^{-1}\ s^{-2}$
Volt	V	$J\ C^{-1}$	$kg\ m^2\ s^{-3}\ A^{-1}$

Appendix B: Physical Constants

These are given to 4 significant figures. There will be cases where you will need to find more precise values.

Constant	Symbol	Value
Avogadro constant	L	6.022×10^{23} mol^{-1}
Bohr radius	a_o	5.292×10^{-11} m
Boltzmann constant	k	1.381×10^{-23} J K^{-1}
Electronic charge	e	1.602×10^{-19} C
Electron rest mass	m_e	9.109×10^{-31} kg
Faraday constant	F	9.649×10^{4} C mol^{-1}
Gas constant	R	8.314 J K^{-1} mol^{-1}
Planck constant	h	6.626×10^{-34} J s
Velocity of light in a vacuum	c	2.998×10^{8} m s^{-1}

Answers to exercises

Chapter 1

1. (a) 2   (b) −5   (c) −6

2. (a) −2   (b) 7   (c) −5

3. (a) 28   (b) −12   (c) 21

4. (a) 7   (b) −7   (c) 9

5. (a) 9   (b) −1   (c) 7

6. (a) $y = x - 10$ (b) $y = 8 - 3x$ (c) $y = 2x - 5$

7. (a) 3^6 (b) x^6 (c) x

8. (a) 6^{-1} (b) x^{-1} (c) y^6

9. (a) 4^{12} (b) x^{12} (c) y^{-12}

10. (a) 3^2 (b) $x^{\frac{5}{3}}$ (c) $y^{-\frac{3}{5}}$

11. (a) 4^{-3} (b) $3x^{-5}$ (c) $2y^2$

12. (a) $\dfrac{4}{5}$ (b) $\dfrac{2}{3}$ (c) $\dfrac{2x}{3y}$

13. (a) $\dfrac{37}{28}$ (b) $\dfrac{13}{8}$ (c) $\dfrac{1}{9}$

14. (a) $\dfrac{21}{32}$ (b) $\dfrac{16}{27}$ (c) $\dfrac{5}{21}$

15. (a) $\dfrac{9}{10}$ (b) $\dfrac{3}{4}$ (c) $\dfrac{9}{10}$

16. (a) 3.6387×10^4 (b) $2.238\,000 \times 10^6$ (c) 1.76×10^{-4}

17. (a) 36 200 000   (b) 16 600   (c) 0.000 012 7

18. (a) $-\dfrac{3}{16}$ (b) $\dfrac{127}{72}$ (c) $-\dfrac{8}{9}$

19. (a) −1   (b) 32   (c) −117

20. (a) $\dfrac{5}{24}$ (b) $-\dfrac{67}{48}$ (c) $\dfrac{41}{72}$

Chapter 2

1. (a) 350   (b) 0.0012   (c) 0.54   (d) 13

2. (a) 13.85   (b) 0.25   (c) 99.54   (d) 0.01

3. (a) 78 500   (b) 0.006 75   (c) 12.0   (d) 80 000

4. (a) 3.1   (b) 12.8   (c) 0.0   (d) 1.0

5. (a) 15.6   (b) 19.1   (c) 2.3   (d) 2.2   (e) 13.9

6. (a) 65.48   (b) 1160   (c) 1.24

7. (a) 10.5   (b) 4.4   (c) 1.72

8. (a) 2.85   (b) 0.36   (c) 3.62

9. (a) 0.4   (b) 0.01   (c) 1%

10. (a) 186 cm^3 (b) 0.017   (c) 1.7%

11. (a) 108 cm   (b) 0.010   (c) 1.0%

12. (a) 0.14   (b) 0.34   (c) 0.10

13. (a) 0.020   (b) 0.064   (c) 1.0×10^{-3}

14. 31.2 ± 0.4 m s^{-1}

15. (a) 10.63   (b) 10.62   (c) 10.61

16. 0.17

17. 0.281

18. 0.08

19. 0.38

20. 0.04

Chapter 3

1. (a) $y = 7 - x$ (b) $y = 3x - 6$ (c) $y = \dfrac{12}{x}$

2. (a) $y = \dfrac{x}{1 - x}$ (b) $y = \dfrac{2}{\sqrt{x}}$ (c) $y = \dfrac{5}{x}$

3. (a) $x^3y + xy^3$ (b) $x^3 + x^2y - x^2y^2 - xy^3$ (c) $x^2 - y^2$

4. (a) (i) 4   (ii) 228   (iii) $12x^3 + 6x^2 + 8x + 1$

  (b) (i) 6   (ii) 5589   (iii) $48x^5 + 5x^4$

  (c) (i) 3   (ii) −215   (iii) $24x^2$

5. $-\dfrac{1}{x^2} + \dfrac{4}{x^3} - \dfrac{9}{x^4} + \dfrac{16}{x^5} - \dfrac{25}{x^6}$

6. (a) 1.380   (b) 1   (c) 1.099

7. (a) gradient = 6, intercept = 3

  (b) gradient = $\dfrac{5}{3}$, intercept = $\dfrac{4}{3}$

  (c) gradient = $\dfrac{1}{2}$, intercept = −4

8.

9.

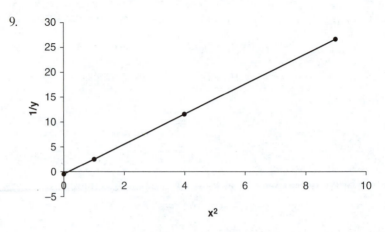

10. 3

11. 1.8

12. $y = \dfrac{64}{x}$

13. (a) −1   (b) 0   (c) 16

14. (a) 0   (b) 0   (c) $\dfrac{3}{2}$

   $f(x,y)$ is not defined when $x = 0$ or $y = 0$.

15. (a) 1   (b) 2.69   (c) 4.10

16. $\dfrac{\partial f}{\partial x} = 6x + 2y + 12x^2 y^2$, $\dfrac{\partial f}{\partial y} = 2x + 8x^3 y$

   $df(x, y) = (6x + 2y + 12x^2 y^2)dx + (2x + 8x^3 y)dy$

17. $\dfrac{\partial^2 f}{\partial x \partial y} = \dfrac{\partial^2 f}{\partial y \partial x} = 2 + 24x^2 y$

18. $2x + 2xy + y^2$, $x^2 + 2xy + 2y$

19. (a) $x = 0.72$ or $x = -9.72$

   (b) $x = -1.39$ or $x = -0.36$

   (c) $x = 0.24$ or $x = 4.27$

20. $x = -1.33$ or $x = 0.5$

Chapter 4

1. 1.03

2. x^2 and y are inversely proportional.

3. $y = 3x^2$

4. $r = \dfrac{Kxy}{z^2}$

5. $x = 4.327$

6. $x = 16$

7. (a) y against $\ln x$, gradient = 6, intercept = 4

   (b) $\log y$ against $\sqrt{x}$, gradient = 2, intercept = 0

   (c) $\ln y$ against $\log x$, gradient = 2, intercept = -3

8. (a) t against s^2, gradient = 1, intercept = -5

   (b) w against v, gradient = 3, intercept = 9

   (c) a against b, gradient = 1, intercept = 14

9. (a) y against $\dfrac{1}{x}$, gradient = 4, intercept = 7

   (b) $\log 2y$ against $\dfrac{1}{x^2}$, gradient = 1, intercept = 6

   (c) $\dfrac{1}{y^2}$ against $\dfrac{1}{x}$, gradient = 1, intercept = ½

10. $y = 1.8x - 0.4$

11. $a = 4.2, b = -3.8$

12. (a) $x = -3$ and $x = 3$ \qquad (b) $x = -\dfrac{3}{2}$

13. (a) $x = 0$ and $y = 0$ \qquad (b) $x = 0$ or $x = 1$

14. $x = 1, y = 0$

15. $\dfrac{df(x)}{dx} = 45x^4 + 9x^2 - 2$

    $\dfrac{d^2 f(x)}{dx^2} = 180x^3 + 18x$

16. $4 - \dfrac{6}{y^3}$

17. $x = \pm 2$

18. $x = \frac{1}{3}$, minimum

19. $x = 0.397$

20. Point of inflexion at $x = 1$

Chapter 5

1. 1.9

2. (a) $\dfrac{x^6}{6} + C$ (b) $x^7 + C$ (c) $-\dfrac{1}{3x^3} + C$

3. (a) $2x^3 + \dfrac{9}{2}x^2 + 8x + C$ (b) $\dfrac{3}{4}x^4 + \dfrac{4}{3}x^3 + C$

    (c) $\dfrac{3}{5}x^5 - \dfrac{4}{3}x^3 + 6x + C$

4. (a) $\dfrac{2}{3}$ (b) 48   (c) 28 068

5. (a) $\dfrac{1}{2}x^2 + \ln x + C$ (b) 6.746   (c) 2425.8

6. (a) $-\dfrac{1}{2x^2} + C$ (b) 0.7997   (c) -6.71×10^{-5}

7. (a) $x^3 + 5\ln x - \dfrac{1}{x^2} + C$ (b) 666 620.77

    (c) $\dfrac{9}{8}x^8 - \dfrac{4}{3}\ln(3x - 1) + \dfrac{3}{5x^5} + C$

8. (a) log 12   (b) ln 6   (c) ln 2

9. $\dfrac{x}{(x + 2)(x + 3)} = \left(\dfrac{3}{x + 3}\right) - \left(\dfrac{2}{x + 2}\right)$

10. (a) $4x + \dfrac{3}{x}$ (b) $3 + \dfrac{1}{x - 2}$

    (c) $12x^2 - \dfrac{2}{2x + 1}$

11. (a) $\dfrac{x^2}{2} + \ln(x - 2) + C$ (b) 0.444   (c) -1.262

12. 0.067

13. (a) 0.0996   (b) 29.56   (c) 3

14. (a) e^6 (b) e^9 (c) x^2e^3

15. (a) $x = 1.623$ (b) $x = -0.288$ (c) $x = 1.082$

16. (a) $x = 36.6$ (b) $x = 335$ (c) $x = 62\ 946\ 1$

17. (a) $arcf(x) = \sqrt{\dfrac{x-5}{3}}$ (b) $arcf(x) = \left(\dfrac{x}{2}\right)^2 + 8$

     (c) $arcf(x) = \sqrt{\dfrac{7}{8(x-2)}}$

18. $x = 6.5, y = -10.5$

19. $x = -2, y = 3$

20. $x = 2, y = -2, z = 0$

Chapter 6

1. 9.51 cm

2. (a) 2.8   (b) 5.0   (c) 7.6

3. (a) 3.0   (b) 2.5   (c) 9.5

4. 0.82, 0.59, 1.38

5. 34.9°, 55.1°, 90.0°

6. (a) 0.191   (b) 0.848   (c) –3.078

7. (a) 0.8660   (b) –0.7071   (c) 0.5774

8. (a) $x = 20.6°$ (b) $x = 96.2°$ (c) $x = 37.4°$

9. (a) $x = -0.280$ rad   (b) $x = -0.343$ rad   (c) $x = 1.585$ rad

10. (a) 4.1   (b) 7.5   (c) 6.2

11. (a) $0.53\,\mathbf{i} - 0.27\mathbf{j} + 0.80\mathbf{k}$

     (b) $-0.14\mathbf{i} - 0.27\mathbf{j} + 0.95\mathbf{k}$

     (c) $0.30\mathbf{i} - 0.90\mathbf{j} + 0.30\mathbf{k}$

12. (a) $3\mathbf{i} + \mathbf{j} - 2\mathbf{k}$ (b) $-\mathbf{i} + 3\mathbf{j} - 4\mathbf{k}$ (c) -3 (d) $-\mathbf{i} - 7\mathbf{j} - 5\mathbf{k}$

13. $105.8°$

14. (a) $4, 2$ (b) $1, -8$ (c) $3, -2$

15. (a) $7 + 3i$ (b) $8 + 2i$ (c) $2 - 12i$

16. (a) $17 + i$ (b) $-1 + 5i$ (c) $24 - 10i$

17. $\cos kx = \dfrac{e^{ikx} + e^{-ikx}}{2}$ $\sin kx = \dfrac{e^{ikx} - e^{-ikx}}{2i}$

18. (a) $\begin{pmatrix} 3 & 2 \\ 7 & 3 \end{pmatrix}$ (b) $\begin{pmatrix} 1 & 2 \\ -1 & -1 \end{pmatrix}$ (c) $\begin{pmatrix} 10 & 4 \\ 7 & 2 \end{pmatrix}$

19. $\begin{pmatrix} 7 \\ 10 \\ 6 \end{pmatrix}$

20. $\begin{pmatrix} 32 & 22 \\ 31 & 32 \end{pmatrix}$

Chapter 7

1. (a) $9.006\,53$ (b) $8.204\,707\,47$ (c) $7.666\,854$ (d) $17.760\,919\,33$

2. $xy = 10.11$

3. (a) 611 (b) $0.007\,35$

4. Minimum at $x = -0.135$

5. Minimum in $f(x)$ at $(1,6)$, minimum in $g(x)$ at $(1,0)$

6. $\dfrac{15}{3}$

7. 12.76

8. (a) $4 - 6i$ (b) $-2 + i$ (c) $9 + 4i$

9. $-12 - 66i$

10. $2, 12, 30, 56, 90$

11. (a) 16, 19   (b) 37, 50   (c) 26, 51

12. $f(n) = 3 \times 2^n$ for $n \geq 0$

13. $arcf(x) = \dfrac{\sqrt{x} - 1}{2}$

14. $arcf(x) = \sqrt{\dfrac{1}{a} \ln\left(\dfrac{x}{2}\right)}$

15. (a) $0.3x^{-0.7}$ (b) $-1.2x^{-1.6}$ (c) $6y^{-2.5}$ (d) $-0.25t^{-1.5}$

16. (a) $2x \ln x + x$ (b) $e^{2x}(2x + 1)$ (c) $2y\, e^{-4y} - 4y^2\, e^{-4y}$

17. (a) $2e^{2x}$ (b) $-3e^{-3x}$ (c) $3 \exp(3x + 2)$

18. (a) $\dfrac{e^{4x}}{4} + C$ (b) $\dfrac{e^6 - 1}{2}$ (c) $\dfrac{e^6 - e^{-12}}{6}$

19. $\dfrac{(e^2 + 1)}{4} = 2.097$

20. $x \ln x - x + C$

Chapter 8

1. 6.6, 5.3

2. 0.380

3. −4.1

4. (a) x is inversely proportional to $\ln y$

    (b) x^2 is directly proportional to y

    (c) x is inversely proportional to e^{-y}

5. (a) y is directly proportional to x , constant $= \dfrac{a}{b}$

    (b) y is inversely proportional to x^2, constant $= a^2 b$

    (c) y is directly proportional to $\ln x$, constant $= \dfrac{a}{b}$

6. (a) $\ln x$ (b) $\ln x^3 y$ (c) $\ln\left(\dfrac{1}{y^2}\right)$

7. (a) $\frac{1}{2}\ln (x + 1) - \frac{1}{2}\ln (x - 1)$

    (b) $\ln 5 + 4 \ln x$

    (c) $\ln 3 + 2 \ln x + \ln y - \ln z$

8. (a) $\log 2$ (b) $\log x$ (c) $\log 5x^2$

9. 31

10. $2N + 3$

11. 2

12. (a) 0   (b) $\cos x = b + 2cx$ (c) 1   (d) 0   (e) $\sin x = x$

13. (a) $x^4 + 12 x^3 + 54 x^2 + 108 x + 81$

    (b) $32 x^5 - 320 x^4 + 1280 x^3 - 2560 x^2 + 2560 x - 1024$

    (c) $729 x^6 + 2916 x^5 + 4860 x^4 + 4320 x^3 + 2160 x^2 + 576 x + 64$

14. (a) $x^5 + 5x^4y + 10x^3y^2 + 10x^2y^3 + 5xy^4 + y^5$

    (b) $\dfrac{81}{x^4} - \dfrac{108y}{x^3} + \dfrac{54y^2}{x^2} - \dfrac{12y^3}{x} + y^4$

    (c) $16x^4 + 160x^3y + 600x^2y^2 + 1000xy^3 + 625y^4$

15. -108

16. 0.988 060

17. $f(n) = 2^{n-1}$ for $n \geq 1$

18. Write Pascal's triangle in form

    1
    1 1
    1 2 1
    1 3 3 1
    1 4 6 4 1

    Diagonals are then 1, 1 + 1, 2 + 1, 1 + 3 + 1, 1 + 4 + 3, etc.

19. $\dfrac{-6i \sin \pi y}{2\pi(9 - y^2)}$

20. $-\dfrac{1}{2\pi iy}e^{-iy} + \dfrac{1}{\pi y^2}e^{-iy} + \dfrac{1}{\pi iy^3}e^{-iy} - \dfrac{1}{\pi iy^3}$

Chapter 9

1. 18

2. 1716

3. (a) 840   (b) 0.05   (c) 75 600

4. (a) n (b) $n(n-1)(n-2)(n-3)$ (c) $(n+2)(n+1)n$

5. (a) $\dfrac{6!}{3!}$ (b) $\dfrac{13!}{5!11!}$ (c) $\dfrac{2!9!}{(6!)^2}$

6. (a) $\dfrac{n!}{(n-3)!}$ (b) $\dfrac{(n+1)!}{(n-2)!}$ (c) $\dfrac{(n+6)!}{(n+2)!}$

7. (a) 8(6!)   (b) 1443(7!)   (c) 89(8!)

8. (a) $(n-1)(n-1)!$ (b) $(n^2-n-1)(n-2)!$ (c) $(n^2+n-1)(n-1)!$

10. $\dfrac{n}{mx}$

11. (a) $2x\ln x + x$ (b) $(3x^2+2x^3)e^{2x}$ (c) $e^{-4x}(1-4x)$

12. (a) $\sin x + x\cos x$ (b) $2x\cos x - x^2\sin x$

     (c) $e^x\cos(3x+1) - 3e^x\sin(3x+1)$

13. (a) 24.2   (b) $\dfrac{1}{3}(x^2+2)^{\frac{3}{2}} + C$ (c) 539

14. (a) $\dfrac{2}{3}(e^x+2)^{\frac{3}{2}} + C$ (b) 78   (c) $\dfrac{(\ln x)^2}{2} + C$

15. (a) $\ln(x-5) + C$ (b) 0.373   (c) $\dfrac{5}{2}\ln(2x+1) + C$

16. (a) −0.165   (b) $-\cos(x\text{-}5) + C$ (c) 0.0

17. 385

18. 1.198 862

19. 872

20. 797 148

Answers to problems

Chapter 1

1. $0 \, \text{kJ mol}^{-1}$

2. $V(r) = -\dfrac{Ze^2}{4\pi\varepsilon_o r}$

3. $T = \dfrac{\Delta H - \Delta G}{\Delta S}$

4. (a) $K = \dfrac{[\text{CH}_4]}{[\text{CO}][\text{H}_2]^2}$

   (b) $K = \dfrac{[\text{CH}_4][\text{H}_2\text{O}]}{8[\text{CO}]^4} = \dfrac{2[\text{CH}_4][\text{H}_2\text{O}]}{[\text{H}_2]^4}$

   (c) $K = \dfrac{8[\text{CH}_4][\text{H}_2\text{O}]}{[\text{CO}]^4} = \dfrac{[\text{CH}_4][\text{H}_2\text{O}]}{2[\text{H}_2]^4}$

   (d) $K = \dfrac{[\text{CH}_4][\text{H}_2\text{O}]}{[\text{H}_2]^4} = \dfrac{[\text{CH}_4][\text{H}_2\text{O}]}{[\text{CO}]^4}$

5. (a) $kt = \dfrac{1}{[\text{NO}_2]} - \dfrac{1}{[\text{NO}_2]_o}$

   (b) $kt = \dfrac{[\text{NO}_2]_o - [\text{NO}_2]}{[\text{NO}_2][\text{NO}_2]_o}$

6. $\dfrac{1}{\phi} = \dfrac{(k_\text{F} + k_\text{IC} + k_\text{ISC})(k_\text{R}[\text{R}] + k_\text{P} + k'_\text{ISC})}{k_\text{ISC}k_\text{R}[\text{R}]}$

7. (a) $c = 2.998\,000\,00 \times 10^8 \, \text{m s}^{-1}$

   (b) $R = 1.097\,000\,0 \times 10^7 \, \text{m}^{-1}$

   (c) $F = 9.649\,0 \times 10^4 \, \text{C mol}^{-1}$

   (d) $p = 1.013\,25 \times 10^5 \, \text{Pa}$

   (e) $l = 1.609\,244 \times 10^6 \, \text{mm}$

8. (a) $\tilde{v} = \dfrac{3RZ^2}{4}$

    (b) $\tilde{v} = \dfrac{5RZ^2}{36}$

    (c) $\tilde{v} = \dfrac{3RZ^2}{4n_1^2}$

9. $10.32 \text{ J K}^{-1} \text{ mol}^{-1}$

10. (a) $v = k[O_3]^{\frac{4}{3}} = k[N_2O_5]^{\frac{4}{3}}$

    (b) $v = \dfrac{k[O_3]^{\frac{4}{3}}}{2^{\frac{2}{3}}} = 2^{\frac{2}{3}}[N_2O_5]^{\frac{4}{3}}$

    (c) $v = 0.072 \text{ mol}^{\frac{4}{3}} \text{ dm}^{-4}$

Chapter 2

1. (a) (i) 1.54 Å   (ii) 1.54 Å

    (b) (i) 25.0°C   (ii) 25.01°C

    (c) (i) -433 kJ mol^{-1} (ii) $-432.88 \text{ kJ mol}^{-1}$

    (d) (i) 16.0   (ii) 16.00

    (e) (i) $8.31 \text{ J K}^{-1} \text{ mol}^{-1}$ (ii) $8.31 \text{ J K}^{-1} \text{ mol}^{-1}$

2. (a) 2.8 g cm^{-3} (b) 15.03 g   (c) 159 kJ mol^{-1}

    (d) 7.9 kJ   (e) $0.042 \text{ mol dm}^{-3}$

3. (a) -283 kJ mol^{-1} (b) $35.21 \text{ J K}^{-1} \text{ mol}^{-1}$ (c) $133 \text{ dm}^3 \text{ mol}^{-1}$

4. 1.0012

5. $0.12 \text{ J K}^{-1} \text{ mol}^{-1}$

6. $5.64 \pm 0.24 \text{ m}^3$

7. $3.07 \pm 0.05 \%$

8. mean = median = 494 kJ mol^{-1}, mode = 493 kJ mol^{-1},

    variance = $4 \text{ (kJ mol}^{-1})^2$, standard deviation = 2 kJ mol^{-1}

9. 0.007 Å

10. 0.001 Å

Chapter 3

1. (a) $\Delta S = \dfrac{\Delta H - \Delta G}{T}$ (b) $V_1 = \dfrac{V - n_2 V_2}{n_1}$ (c) $n = \dfrac{a}{h}\sqrt{8mE}$

2. (a) $\dfrac{nR}{V}$ (b) $\dfrac{RT}{V}$ (c) $\dfrac{RT}{p}$

3. Plot C_p against T: gradient = b, intercept = a

4.

5. $\dfrac{\Delta H^o}{RT^2}$

6. $\dfrac{dCp}{dT} = b - \dfrac{2c}{T^3}$

7. $\left(\dfrac{nR}{p}\right)dT - \left(\dfrac{nRT}{p^2}\right)dp$

8. $a = 0.138$ Pa m^6 mol^{-2}, $b = 3.18 \times 10^{-5}$ m^3 mol^{-1}

9. $V = \dfrac{RT \pm \sqrt{R^2T^2 - 4ap}}{2p}$

10. -0.32

Chapter 4

1. $0.274 \text{ mol kg}^{-1}$

2. (a) π against c, gradient $= RT$, intercept $= 0$

   (b) $\log k$ against $\sqrt{I}$, gradient $= 1.02\, z_A\, z_B$, intercept $= \log k_0$

   (c) Λ against $\sqrt{c}$, gradient $= -(P + Q\,\Lambda_0)$, intercept $= \Lambda_0$

3. (a) Plot E against $\ln Q$, gradient $= -\dfrac{RT}{zF}$, intercept $= E^\circ$

   (b) Plot E against T, gradient $= -\dfrac{R}{zF}\ln Q$, intercept $= E^\circ$

4. $0.325 \text{ mol dm}^{-3}$

5. $\dfrac{\partial c}{\partial t} = D\,\dfrac{\partial^2 c}{\partial x^2}$

6. (a) $V_1 = \dfrac{1}{\rho}$ \qquad (b) $\Delta T_b = \dfrac{K_b m}{2}$

7. (a) Continuous, because $v > 1$

   (b) $\alpha = 1$, complete dissociation

8. $17.6 \text{ cm}^3 \text{ mol}^{-1}$

9. 1005.65 cm^3

10. Minimum at $x = 0.433$.

Chapter 5

1. $4.6 \times 10^{-10} \text{ mol dm}^{-3} \text{ s}^{-1}$

2. $\dfrac{d[NH_4Cl]}{dt} = \dfrac{1}{3}\dfrac{d[Cl_2]}{dt} = -\dfrac{d[NCl_3]}{dt} = -\dfrac{1}{4}\dfrac{d[HCl]}{dt}$

3. $\dfrac{1}{1-a}\left([A]^{1-a} - [A]_o^{1-a}\right) = -kt$, not valid when $a = 1$.

4. $\log v = \log k + m \log [NO] + n \log [Cl_2]$

   Keep concentration of one reactant constant. Plot $\log v$ against $\log c$ where $c =$ concentration of other reactant. Order (m or n) is given by gradient.

5. First order, $k = 3.66 \times 10^{-3}$ s^{-1}.

6. 6.3×10^3 s^{-1}

7. 81 kJ mol^{-1}

8. 190 kJ mol^{-1}

9. $[Br] = \sqrt{\dfrac{k_1}{k_5}[Br_2]}$

10. $[NO] = \dfrac{k_2[NO_2]}{k_3}$

Chapter 6

1. $\dfrac{\pi}{6}$

2. 1.47 Å

3. 1.922 Å

4. 3.150 Å

5. 36.5°

6. 110.2°

7. $a\mathbf{i}$, $a\mathbf{j}$, $a\mathbf{k}$, ½$a\mathbf{i}$ + ½$a\mathbf{j}$ + ½$a\mathbf{k}$

8. 364.3 Å^3

9. (a) $F(0\ 0\ l) = f_1 \exp 2\pi i l z + f_2 \exp 2\pi i l(z + 0.5)$

10. $\begin{pmatrix} x_2 \\ y_2 \end{pmatrix} = \begin{pmatrix} 1 & 0 \\ 0 & -1 \end{pmatrix} \begin{pmatrix} x_1 \\ y_1 \end{pmatrix}$

Chapter 7

1. $5.291\ 772\ 6 \times 10^{-11}$ m

2. Minimum at $r = \sqrt[3]{\dfrac{3A}{2B}}$

3. $arcE(n) = \dfrac{2d}{h}\sqrt{mn}$

4. $\nu - \nu_o = 2.52 \times 10^{14}\ T_{max}$ where $\nu_o = 5.47 \times 10^{14}$ Hz

5. $arc\Psi_{1s}(r) = -\left(\dfrac{a_o}{Z}\right) \ln\left[\dfrac{r}{2}\left(\dfrac{a_o}{Z}\right)^{\frac{3}{2}}\right]$

6. $\dfrac{dm}{dE} = \dfrac{\pi}{h}\left(\dfrac{2I}{E}\right)^{\frac{1}{2}}$

7. $A = \dfrac{1}{\sqrt{2\pi}}$

8. $\dfrac{a^3\hbar^2}{6m}$

9. $\left(\dfrac{a}{2\pi n}\right)^2\left(\dfrac{\pi^2 n^2}{3} - 2\right)$

10. $\dfrac{3}{2}a_o$

Chapter 8

1. 4.23 D, 1.74×10^{-39} C m^2

2. 6.949×10^{13} Hz, 4.314×10^{-6} m, 4.604×10^{-20} J

3. $I = I_o e^{-bl}$

4. $F(J+1) - F(J) = 2B\,(J+1) - 4D(J+1)^3$

5. $8.417\,27 \times 10^{-22}$ J

6. $G(1) - G(0) = \omega_e - 2\,\omega_e\,x_e$

7. 3.452×10^{-6} m

8. $3.7955 - 0.035\ m$

9. The peak due to $(CH_3)_2$ will be split into two in the ratio 1:1. The peak due to CHI will be split into seven in the ratio 1:6:15:20:15:6:1.

10. $g(\omega) = \dfrac{1}{2\pi(a+i\omega)}\left[e^{-(a+i\omega)} - \dfrac{e^{-2(a+i\omega)}}{2}\right]$

Chapter 9

1. 30

2. (a) $N!$ (b) N

3. $p = \dfrac{NkT}{V}$

4. (b) $N^N e^{-N}$ (c) 14%

5. $q' = \displaystyle\sum_i g_i \exp\!\left(\dfrac{-(\varepsilon_i + \alpha)}{kT} \right)$

6. 4

7. (a) 2.104 38 × 10⁻²² J   (b) 19.561

8. 8.4325 × 10⁻⁷, 1.4039 × 10⁻⁵

9. $\dfrac{dS}{dT} = \dfrac{12.47}{(T/K)}$

Chemical index

Mathematical index